JAN 1 4 2022

D0890047

JAN 1 4 2022

# The Steger
# Homestead Kitchen

# The Steger Homestead Kitchen

Will Steger and Rita Mae Steger

~ with Beth Dooley

## Simple Recipes
## for an Abundant Life

University of Minnesota Press
Minneapolis
London

The University of Minnesota Press gratefully acknowledges the generous assistance provided for the publication of this book by the Margaret W. Harmon Fund.

Photographs on pages 15, 20, 42, 49, 52, 59, 73, 77, 78, 96, 113, 115, 116, 137, 139, 149, and 156 copyright Mette Nielsen; pages ii, x, 2, 3, 10, 12, 16, 27, 38, 46, 64, 66, 86, 90, 92, 99, 104, 108, 122, 129, 140, 144, 153, 155, and 160 copyright John Ratzloff; page 6 courtesy Will Steger; page 170 copyright Rita Mae Steger; page xiv copyright Justin Chung; page 5 by Marlin Levison copyright 2013 *Star Tribune*.

Copyright 2022 by Will Steger and Rita Mae Steger

All rights reserved. No part of this publication may be reproduced, stored in a retrieval system, or transmitted, in any form or by any means, electronic, mechanical, photocopying, recording, or otherwise, without the prior written permission of the publisher.

Published by the University of Minnesota Press
111 Third Avenue South, Suite 290
Minneapolis, MN 55401-2520
http://www.upress.umn.edu

ISBN 978-1-5179-0974-1

A Cataloging-in-Publication record for this book is available from the Library of Congress.

Printed in Canada on acid-free paper

The University of Minnesota is an equal-opportunity educator and employer.

28 27 26 25 24 23 22     10 9 8 7 6 5 4 3 2 1

*This book is dedicated to the many people who have worked with me to create The Homestead and share the lessons of this wildly beautiful place. Together we've built cabins, our sauna, and the Lodge—we've raised dogs, founded a wilderness school, explored the Arctic. We have struggled, grieved, and celebrated. I offer these stories and our recipes in my mother's tradition of generosity. May they feed friendships, understanding, and gratitude.*

—WILL STEGER

# Contents

## 3. Homestead Gatherings

## 4. Fireside Feasts

## 5. The Homestead Oven

## 6. Wholesome Treats

## 7. The Pantry

# The Steger Homestead Kitchen

*Beth Dooley*

I'm neither homesteader nor Arctic explorer, but I am a fan of Will Steger's work, captivated by stories of his thrilling expeditions and his ability to live so well with so little. A few years ago, at the invitation of mutual friends John Ratzloff (who took many of the beautiful photographs that fill this book) and landscape designer Kim Knutson, I visited the Steger Wilderness Center in northern Minnesota, eager to learn about Will's no-waste, climate-friendly kitchen.

It was in August and my first evening, after a sauna and a swim in Picketts Lake, I hiked from my floating sleeping cabin to the lodge where The Homestead cook Rita Mae Steger served our motley crew a wonderful dinner on the deck. Gathered were an electrician, a stonemason, a plumber, several interns, apprentices, students, a poet, and a photographer. We filled our plates with vibrant curry, a garden salad, and just-picked raspberry pie, then lingered near a bonfire as loon calls echoed over the water.

Rita Mae, Will's energetic niece, rustles up three meals a day at The Homestead, with fresh ingredients spiced with the flavors of her Vietnamese mother's cooking and her own travels through Asia. She cooks with creativity and thrift in a very rustic, sparsely equipped kitchen. Like Will she takes on everyday tasks with joy. They are kindred spirits of different generations.

To create this book, Rita Mae and I gathered and tested recipes—hers, mine, and those from Will's family and friends that reflect The Homestead's values. Woven through these pages are Will's stories of this home, of the people he has met on his expeditions, and of those drawn to this place. In working with Will and Rita Mae, I learned that homesteading is about more than clearing land and building structures: it is a practical and necessary approach to living lightly on the earth. Homesteading is not about taming wilderness; it's about garnering our resources. Food is not just a list of ingredients—it's pasture-raised beef, pork, chicken, dairy, and eggs from farmers nearby, it's seasonal vegetables like rhubarb that return year after year. It's the grains and beans grown on regenerative farms.

You'll find lots of healthy plant-based recipes in these pages, along with plenty of pasture-raised meat. The pasture grasses and flowers capture carbon from the atmosphere and return it to the soil where it belongs. Such farms do not need to use toxic chemicals because the animals do the work of fertilizing and nourishing the soil by turning it over with their hooves and providing compost. Keeping the ground covered with plants throughout the year stems water runoff, retains precious topsoil, and provides friendly habitat for pollinators and wildlife. Buying food from regenerative farms also supports farmers and the rural economy. The Homestead kitchen does not purchase meat, poultry, eggs, or dairy products from CAFOs (contained animal feeding operations)—these are factories, not farms. Pasture-raised food may cost more, so the kitchen uses less and wastes nothing.

"We need to live with what we have," Will says. "Yes, recycling is important, but if we stopped buying so much stuff, we wouldn't be throwing so much out." Working with Rita Mae and Will has been humbling and inspiring. Through these recipes you'll get to know Rita Mae and understand her approach to sourcing and cooking for The Homestead.

Like all of Will's initiatives, this book is a collaborative effort of many family members and friends—and now you are one, too. Welcome to The Homestead.

# The Steger
# Homestead Kitchen

# An Arctic Explorer at Home

*Will Steger*

Most know me as a polar explorer, but for much of my life I have been a designer and builder who has made a home in the wilderness. Fifty years ago, I moved out of Minneapolis to eke out an independent homestead in the remote roadless area north of Ely, Minnesota. I built my log cabin and a sauna, then settled into the daunting task of clearing the rocky terrain by hand for my gardens and orchard. The Homestead now encompasses two hundred acres and is the site of the Steger Wilderness Center, with a stand-alone microgrid that powers the full campus including the convening center, Lodge, workshop, sleeping cabins, root cellars, and icehouse.

Establishing The Homestead was not a solo endeavor. I credit the hard physical work and commitment of an engaged community of architects, carpenters, stonemasons, electricians, interns, apprentices, high school students, gardeners, cooks, and my extended family and friends. We felled trees, sawed lumber, paved paths, built stone walls, cut and hauled ice, cooked and dined together. We gathered around an open fire in the summer and the woodstove in the winter. The Homestead was built with a lot of love.

The heart of The Homestead has always been the Lodge and kitchen. In the summer, a good cook is essential to feeding the hungry crew delicious, nutritious, comforting meals. I couldn't have found anyone better for that substantial task than Rita Mae Steger, my niece. She is a second-generation homesteader, the daughter of my brother Bob, who came of age at The Homestead in the rollicking 1960s. He introduced Rita Mae to The Homestead when she was a toddler, and she has come back every summer for twenty-five years.

Rita Mae is dauntless, curious, and independent. Rita Mae's mom, Kim Chi, is Vietnamese, and Rita Mae spent much of her childhood visiting her relatives in the rural regions of that beautiful country. She has traveled extensively throughout Asia and is an explorer at heart. She also

understands that her role at The Homestead goes well beyond simmering fragrant stews and flipping pancakes: she mentors the young women who work here, engages the crew in kitchen and garden tasks, but more, she listens and pays attention. Thanks to Rita Mae we all linger at the table long after the plates are cleared, chatting and laughing. Like her mother and grandmothers, she understands that dining together nourishes friendships and understanding.

## The Homestead Story

Even as a young child growing up in the exurbs of St. Paul, I knew I wanted to homestead and carve out a life in the woods. Captivated by stories of my immigrant grandparents who cleared land and farmed, I planted vegetables in my mother's flower gardens and slept in a tent in the backyard to be near them. As the second son in a family of nine, I learned to be independent and get along.

Dinner was a big deal in our house, and God help you if you were late to the table. More than any particular dish or favorite dessert, what I remember are the stories we told each other, the shared tragedies and the triumphs, the games lost and the scholarships won, our conversations while doing chores and washing dishes. Our kitchen featured one of those cafeteria-style milk machines that drew friends to hang out after school. My mom kept the cookie jar full and always added an extra chair at the table to welcome a schoolmate or an associate of my Dad. We didn't live lavishly, but good food was abundant and served with generosity and care.

As a teen, I spent my summers working on my aunt and uncle's farm in southern Minnesota, helping harvest corn and wheat, milking cows, moving the cattle through paddocks, and sleeping in the hayloft, where stars peeped through the rafters. I loved the long days of hard outdoor work, the big meals around the trestle table, the discipline and companionship.

At age nineteen, I struck out on a 3,000-mile kayak adventure from the southern Canadian Rockies to Alaska and was inspired on this trip to settle in northern Minnesota. As soon as I returned, I bought the original twenty-eight acres on which The Homestead sits. I put down a twenty-five-dollar traveler's check and a five-dollar bill, and paid off the rest over time working factory jobs and on truck farms. The property was exactly what I was looking for: three miles and two lakes from the nearest road, on the edge of the Boundary Waters.

My plan was to complete my formal education in the city and then move full-time to the wilderness. I graduated with a degree in geology and biology and a teaching certificate from the University of St. Thomas, then taught junior high science in Minneapolis for three years while I earned a master's degree in education. In the summer of 1970, I hitchhiked north to Alaska to fight forest fires, and I returned in the fall with a grubstake that enabled me to make my pitch in the wilderness.

The early Homestead was rough; the country up north is rugged, with winters of numbing cold and deep snow and summers of merciless bugs — exactly what I had dreamed of. I wanted to build a community in this isolated setting but had no idea how to make a living. I'd heard about a new school in the area called Outward Bound, so I hitchhiked there one autumn Sunday to check the place out. It was my good fortune that during that coming December they would be launching their new winter program, and I was hired on the spot as one of the four instructors. This opened up the paradigm of outdoor education for me.

In the Outward Bound program, students hauled their own gear. I knew that I could create a different and more interesting experience with sled dogs, something no other school was doing at the time. After two years with Outward Bound, I founded my own winter school that used dog teams. I had never worked with dog teams before, so recruiting and purchasing an outfit was a formidable challenge.

Patti Steger, my closest friend and former wife, and I raised her two young sons in our 400-square-foot log cabin. We relied on kerosene for light and wood for warmth and cooking. We lived on our garden's harvest, wild foods foraged in the woods, and the deer we hunted and fish we caught. When the winter school was in session, we bought pork and beef from a local farmer, which we kept frozen in the unheated trip-storage building. Good food and storytelling around blazing fires in the evening were the matrix of our growing wilderness community.

Within four years the gardens were producing a bounty and work began on the root cellar and icehouse to store our food throughout the year. In the early 1970s the co-op system in Minneapolis was on its feet and we were able to buy organic grains and beans in bulk. At that time, a hundred pounds of organic wheat sold for just thirty dollars. Each fall we bought our year's supply of staples from the co-op warehouse, saving us lots of money and at the same time cutting back on our garbage stream. During the first twelve years of The Homestead, we ate all organic food and lived on two thousand dollars a year while generating only four cubic yards of waste.

By the late 1970s The Homestead was settled, and my attention shifted further north. By trial and a lot of error I was by then breeding a strong line of northern dogs that we called Polar Huskies. These dogs eventually opened up the Arctic and polar regions for international teams of explorers that I would lead over the next thirty years.

Sled dogs made my wilderness life complete, providing me with a livelihood at The Homestead. I taught wilderness skills to support myself. The dogs helped me transport massive amounts of lumber, sand, and concrete across the lakes that separated us from the nearest road. The winter school attracted new friends and associates, forming the foundation of our extended community today.

During my years of polar expeditions, I was graced with the radical hospitality of Native elders and trappers along the Arctic coasts and the Dene hunters in the boreal forests. They lived with so little by our standards

and yet insisted on inviting me into their homes to share their food. Even when we didn't speak the same language, dining together fueled friendships, understanding, and gratitude. The role of good food and a balanced diet for both people and dogs played a pivotal role in our successes.

Through my expeditions, I've been eyewitness to the changing climate and topography of the polar regions and keenly aware of the devastating impact of climate change. In 2002, when the Larsen B Ice Shelf in Antarctica collapsed, I was stunned, heartbroken: it felt personal. That magnificent glacial area spanned 1,250 miles and had taken my team two weeks to cross by dogsled. Its rapid disintegration was a disaster on an epic scale that eluded the media. I knew, in that moment, I needed to do more than lecture. I knew then that it was time to leverage into action the attention I was getting for my international expeditions. This was the push I needed to found Climate Generation. This nonprofit organization, now led by Nicole Rom, educates and engages communities in climate solutions. I'm so proud of its work over the past ten years developing curriculum for educators, engaging youth and communities in action, and lobbying policy makers.

My work on climate continues at the Steger Wilderness Center for Innovation and Leadership, a nonprofit based at The Homestead. The winter school helped me become a wilderness educator, and Climate Generation extends my vocation by engaging communities in climate literacy and action. The Wilderness Center is the actual place for both visioning and practical application of this mission. It's the place to train tradespeople in environmentally responsible building skills. To complete the Center's buildings and infrastructure, we are partnering with nonprofits such as Louis King's Summit Academy on the north side of Minneapolis and Dunwoody College of Technology. Once the work on the facilities is completed, we will draw together our leaders in business, politics, and education to strategize on meaningful ways to address climate change. We'll offer programs for students, interns, and the public. Put simply, the Steger Wilderness Center, founded on the site of my home, provides a holistic education that engages head and heart in this remote, wild, and beautiful place. Come on in. Let's cook!

That first year, clearing the land to put in a garden was physically exhausting and exhilarating. I clearly remember one blustery October afternoon. I was digging out roots and rocks and stopped to lean on my shovel for a break. I could hear the snow geese approaching from the far north. Within minutes I looked up and saw them fly in V formations overhead. How many generations had witnessed the snow geese as they readied their homesteads? I felt deeply connected to my ancestors, who had done this work before me and who understood their connection to the earth.

It's no wonder that the garden has drawn so many people to this place over the past fifty years. The wilderness attracts like-minded folks, and food cultivates those relationships and creates community. In many ways, gardening is like being on an expedition. You are tied to the weather, to the rain, to the frost. Life is determined by so many unseen factors. You're constantly adapting because you are not always in control, and you learn to pay attention, allowing experience and intuition to guide your decisions. Raising food requires patience and flexibility; the garden's abundance can be a blessing and a curse. One year, our best crop was turnips—and we ate *lots* of turnips at every meal.

At The Homestead, my life assumes a different rhythm than in the city, where I'm governed by the clock. Here I wake at first light and go to bed when it is dark. During the summer the days are long, and the hours are focused on outside work that can be completed only when it is warm—

fixing a roof, building a wall. It's a season of physical activity. But as the weather cools the tempo slows. We harvest the gardens, chop wood for the winter, put up the deer fence around the orchard. Winter is a relief. When the temperature drops below zero degrees, there's really nothing we can do except read, study, write, and work with our hands in the woodshop.

We dug a root cellar to store food that first year of homesteading. It's carved into the hillside opposite the Lodge and kitchen and is kept cool with blocks of ice that are cut and hauled from the lake and stored in sawdust each winter. The root cellar is cool and humid and is especially good for storing fresh produce. It also protects food from freezing solid when it's below zero. Lit by a skylight that also acts as a vent, it's pleasant in the summer, and come winter the sight and scent of fresh vegetables and herbs lift my spirits.

# Red Cabbage and Celery Salad

Serves 4 to 6

*This is a dish of brilliant contrasting colors—magenta and deep green. Use your fingers to massage the oil and vinegar into the cabbage leaves; this helps them become tender. Make the salad a few hours ahead of serving so the flavors have a chance to marry.*

½ medium red cabbage,
    core removed and thinly sliced, 4 cups
3 tablespoons white balsamic vinegar
2 tablespoons olive oil
3 celery stalks, leaves included, thinly sliced
2 tablespoons chopped parsley, plus extra for garnish
2 tablespoons pine nuts
1 teaspoon sesame seeds
Salt and freshly ground black pepper, to taste

Turn the cabbage into a medium bowl. Pour in the vinegar and oil and then, using both hands, massage them into the sliced cabbage so it becomes tender. Toss in the celery, chopped parsley, pine nuts, and sesame seeds and season to taste with salt and pepper. Garnish with a little extra parsley before serving.

# Roasted Squash Salad

Serves 4

*Butternut squash, cubed and roasted in a hot oven, turns velvety and lush with a fine, caramel crust. Roast up big batches of any winter squash for salads and soups, and to toss into pasta, pilafs, and salads like this one.*

1½ pounds butternut squash, cut in half and seeded and peeled
½ cup extra virgin olive oil
Pinch kosher salt
A few grinds black pepper
1 lemon, juice and zest
1 small shallot, minced
2 teaspoons chopped rosemary
1 good-sized bunch (about ½ pound) dinosaur kale,
    spinach, arugula, and/or romaine
½ cup sliced red onion
Pinch red pepper flakes

Preheat the oven to 400 degrees. Cut the squash into ½-inch cubes and turn into a large bowl. Toss with just a little oil to coat, and sprinkle with a pinch of the coarse salt and a few grinds of pepper. Spread out on a baking sheet and roast until the edges are crisped and the squash is tender, shaking the pan occasionally and turning the pieces with a spatula so they crisp on all sides, about 20 to 25 minutes. Remove from the oven and set aside.

While the squash is roasting, make the vinaigrette. Put the lemon zest and juice, shallot, rosemary, and the remaining oil into a jar with a lid and shake until well combined.

Turn the greens into a salad bowl along with the squash and red onion and drizzle in just enough of the dressing to lightly coat. Season to taste with more salt and pepper and a pinch of red pepper flakes.

# Wild Rice Salad with Roasted Corn

Serves 4 to 6

*Will purchases real wild rice from local "ricers" who harvest it in canoes and toast it over an open fire. Hand-harvested wild rice is mottled brown and tan, and, unlike cultivated black rice, it cooks quickly, in about 15 minutes. We often roast up extra ears of corn (page 93) for this salad.*

2 cups cooked wild rice

1 bunch scallions, white part only, chopped

1 cup steamed or Ember-Roasted Corn kernels (page 93)

¼ cup chopped parsley, plus more for garnish

¼ cup toasted pecans or hazelnuts

2 tablespoons apple cider vinegar

2 tablespoons maple syrup

1 teaspoon Dijon mustard

2 tablespoons nut or vegetable oil

Salt and freshly ground black pepper, to taste

In a medium bowl, toss together rice, scallions, corn, parsley, and nuts. In a small bowl, whisk together vinegar, syrup, mustard, and oil. Season with salt and pepper to taste. Toss just enough of the dressing into the rice mixture to coat the ingredients. Serve at room temperature or cold.

• • •

**TO COOK THE RICE** Put ½ cup of raw wild rice into a colander and rinse under cold running water until the water runs clear. Turn the rice into a deep saucepan and add enough cold water to cover the rice by about 4 inches. Set over high heat, bring to a boil, reduce the heat, and simmer the rice until the grains have opened and are tender but not mushy, about 15 to 20 minutes. Taste, and continue cooking if necessary. Drain.

# Grainy Greek Salad

Serves 4 to 6

*This simple whole grain salad is bright, flavorful, and satisfying, filling you up but not weighing you down.*

2 tablespoons extra virgin olive oil
Juice of 1 small lemon
Salt and freshly ground black pepper, to taste
3 cups cooked brown rice
3 cups spinach
⅔ cup crumbled feta cheese
½ cup pitted, sliced kalamata olives
¼ cup toasted walnuts, chopped

In a small bowl, whisk together the olive oil, lemon juice, and salt and pepper to taste. Turn the rice and spinach into a large bowl and toss in the vinaigrette to coat. Toss in the feta and olives. Serve garnished with the toasted walnuts.

• • •

**TO COOK THE BROWN RICE**  Rinse the rice in a colander under cold running water and turn into a medium pot. Stir in 2 cups of cold water and 1 teaspoon salt. Set over high heat, bring to a boil, then reduce the heat so that the liquid is just barely simmering; cover the pan. Cook the rice for 45 minutes undisturbed. Keep the pot covered, remove from the heat, and let the rice stand for 15 minutes. Fluff the rice before proceeding with the recipe.

**TO TOAST WALNUTS**  Turn the walnuts onto a baking pan and toast in a 350-degree oven until they smell nutty and are crisped and lightly browned, about 8 to 10 minutes. Remove, cool, and chop.

# Farro and Chickpea Salad

Serves 4 to 5

*Several times during the year Will heads into the city and comes back with 100-pound sacks of whole grains and dried beans. They store well, are easy to work with, and provide a big boost of protein at very little cost.*

1 cup farro, rinsed
½ cup chopped parsley
2 cups chopped celery
1½ cups cooked dried or canned chickpeas, drained
¼ cup olive oil
Juice of 1 lemon
Salt and freshly ground black pepper, to taste
Generous pinch red pepper flakes

Put the farro into a pot and add enough water to cover by 4 inches. Set over high heat, bring to a boil, reduce the heat, cover, and simmer until the grain is tender, about 20 to 30 minutes. Drain and set aside.

In a large bowl, toss together parsley, celery, chickpeas, olive oil, and lemon juice. Turn the farro into the bowl and toss together. Season to taste with salt, pepper, and red pepper flakes.

• • •

**TO COOK DRIED CHICKPEAS** Put the chickpeas into a pot and add enough water to cover by 4 inches to soak the chickpeas overnight. Drain in a colander, turn the chickpeas into a large pot, and add enough water to cover by 4 inches. Set the pot over high heat and bring to a boil; reduce the heat to simmer the liquid. Cover and cook the chickpeas until tender, about 45 minutes to an hour, checking to add more water if necessary. Drain the chickpeas and proceed with the recipe.

# Gazpacho

Serves 4 to 6

*We make this on farmers' market days when the heirloom tomatoes are at their peak. I chill it down in the root cellar for a refreshing lunch.*

2 pounds tomatoes, chopped
   (use a mix of different varieties)
1 cup tomato juice or water
2 cloves garlic
2 to 3 slices stale bread,
   crusts removed and torn into pieces
¼ cup extra virgin olive oil, plus more as needed
2 tablespoons red wine vinegar, or to taste
Salt and freshly ground black pepper, to taste
1 small cucumber, chopped
1 medium red bell pepper, seeded and diced
½ cup cooked corn kernels, for garnish (optional)
¼ cup chopped fresh basil, for garnish (optional)

Put the tomatoes, tomato juice, garlic, bread, oil, and vinegar into a blender or roughly chop and stir together in a bowl. Work the ingredients together until smooth. If the mixture seems too thick, add a little more tomato juice or water. Season with salt and pepper to taste. Stir in the cucumbers and pepper. Chill and serve cold. Garnish with more cucumber and pepper, corn, and basil.

# Creamy Creamless Cauliflower Soup

Serves 4

*Quick, easy, light, and creamy without the addition of any dairy, this simple soup is a favorite for lunch, paired with a selection of cold cuts and bread.*

1 tablespoon butter
1 tablespoon unbleached flour
4 cups chicken or vegetable broth
½ cup minced onions
1 medium head cauliflower, chopped, about 4 cups
Salt and freshly ground black pepper, to taste

In a medium saucepan, melt the butter over low heat. Whisk in flour and cook, stirring until the butter mixture is slightly browned and no longer tastes pasty. Stir in the broth, onion, and cauliflower. Bring to a boil, then reduce the heat to low, cover, and simmer until the vegetables are tender, about 15 minutes. Puree the soup in batches with a blender. Return to the pot and warm to the desired temperature, then season with salt and pepper to taste.

# Homey Tomato Soup

Serves 4 to 6

*This vegan version of creamy tomato soup relies on roasted cauliflower, not cream, for flavor and body. Be sure to roast the cauliflower until thoroughly browned for the richest taste.*

I large cauliflower,
   about 2½ to 3 pounds
5 tablespoons olive oil
Coarse salt
I medium onion, thinly sliced
5 cloves garlic, crushed
3 tablespoons tomato paste

½ teaspoon red pepper flakes
5 sprigs fresh thyme
2 cans (28 ounces each)
   crushed tomatoes
I teaspoon salt
I teaspoon sugar
4 cups Vegetable Stock (page 159)

Preheat the oven to 425 degrees. Cut the cauliflower into 2-inch pieces, turn into a medium bowl, and toss with 1 tablespoon of olive oil and a few pinches of coarse salt. Spread out on a large baking sheet so the pieces do not touch. Roast, stirring occasionally, until the cauliflower is browned, about 20 minutes. Remove from the oven and set aside.

Film a deep pot with the remaining oil and set over medium heat. Sauté the onion and garlic until soft, about 2 to 3 minutes. Stir in tomato paste and cook for 1 minute, then add the red pepper flakes, thyme, crushed tomatoes, salt, sugar, and the Vegetable Stock. Bring the liquid to a boil, reduce the heat, and then simmer for 2 to 3 minutes, and then add the roasted cauliflower and continue simmering for another 5 minutes until the cauliflower is very tender. Puree the soup using an immersion blender (or, working in batches, in a blender) until smooth and return the soup to the pot. Serve warm.

# Curried Red Lentil Soup

Serves 4 to 6

*This is a warming, fragrant soup for a gray, chilly day.*
*Because red lentils cook quickly, the soup is ready in less than an hour.*

2 tablespoons coconut oil
1 onion, diced
2 cloves garlic, smashed
1 tablespoon ginger
1 tablespoon curry powder
¼ teaspoon cayenne
1 cup red lentils, rinsed
1 can (15 ounces) crushed tomatoes
1 can (13.5 ounces) unsweetened coconut milk
3 cups water
4 to 6 lime wedges for serving
Chopped cilantro, for garnish

In a large soup pot set over medium heat, melt the coconut oil. Add the onion and sauté until translucent, about 7 minutes. Add the garlic, ginger, curry powder, and cayenne. Cook, stirring frequently, until the spices and aromatics are fragrant, about 2 minutes. Add the rinsed lentils and tomatoes and cook 1 minute. Add the coconut milk and water, bring to a boil, then reduce the heat to a simmer and cook for about 3 minutes. Serve with lime wedge and garnished with cilantro.

# Ginger Squash and Apple Soup

Serves 6

*Butternut squash with smooth, thin skins are the easiest to peel,*
*but any winter squash will work well in this recipe (except spaghetti).*
*Use a tart apple, like Haralson—the variety Will grows in*
*The Homestead orchard.*

2 tablespoons coconut oil or butter

1 small onion, thinly sliced

2 tablespoons thinly sliced fresh ginger,
   or 1 tablespoon dried ginger

2 tablespoons curry powder

1½ pounds butternut or kuri squash,
   peeled and seeded and cut into 1-inch pieces

2½ to 3 cups water

1 can (13.5 ounces) unsweetened coconut milk

1 teaspoon grated lime zest

2 tablespoons fresh lime juice, or more to taste

1 tablespoon honey, or to taste

Salt and freshly ground black pepper, to taste

¼ cup chopped cilantro

In a large heavy pot, heat the coconut oil or butter over moderate heat
and sauté the onion and ginger, stirring until the onion is softened, about
5 to 7 minutes. Stir in curry powder and cook until fragrant, about 1 minute.
Add the squash and enough water to cover and bring to a boil over high heat.
Cover partially, reduce the heat, and simmer over low heat until soft, about
25 minutes. Stir in coconut milk, lime zest and juice, and honey, and season
with salt and pepper. Stir in the cilantro and serve right away.

# Northwoods Mushroom Wild Rice Soup

Serves 4 to 6 (and easily doubled)

*Be sure to use hand-harvested wild rice from the clear lakes
of northern Minnesota and Wisconsin. It cooks in far less time
than the cultivated black paddy rice. The wild mushrooms
enhance the mushroom flavor of the soup.*

1 cup hand-harvested wild rice, rinsed

2 to 3 ounces dried morel mushrooms
(or any dried mushroom)

2 tablespoons unsalted butter

1 stalk celery with leaves, diced

1 medium carrot, diced

1 medium yellow onion, diced

2 cups sliced cremini or shiitake
mushrooms

2 tablespoons chopped parsley,
plus more for garnish

1 tablespoon chopped fresh thyme
or 1 teaspoon dried

Salt and freshly ground black pepper

3 cups reserved stock from cooking
the wild rice or chicken stock

½ cup heavy cream

Splash of sherry or bourbon
(optional)

Generous pinch red pepper flakes,
to taste

Put the wild rice into a colander or sieve and rinse under cold water until
the water runs clear. Turn into a large pot and add enough water to cover the
rice by about 4 inches. Set over high heat, bring to a boil, then reduce the
heat to a simmer and cook just until the rice kernels have opened and are
tender but not mushy, about 15 to 20 minutes. Remove and drain, reserving
the rice cooking water.

Put the dried mushrooms into a medium dish and pour enough of the
rice water over them to cover by about an inch, to rehydrate the mushrooms.

In a 3-quart saucepan set over medium heat, melt the butter and sauté the
celery, carrot, onion, fresh mushrooms, parsley, and thyme with a pinch of

salt and pepper. Cover the pan and "sweat" the vegetables to release their juices, about 3 to 5 minutes. Remove the lid and continue cooking until the vegetables are tender and their juices have evaporated.

Stir in the reserved wild rice stock or chicken stock plus the rehydrated mushrooms and soaking liquid, scraping the bottom of the pan to release any brown nubs. Then add cream, sherry, and wild rice. Season with salt, pepper, and red pepper flakes to taste. Serve garnished with more parsley.

# Garden Borscht with Horseradish Cream

Serves about 6 to 8

*After dinner, Will often shares stories of his polar expeditions.
I'm especially fond of his accounts of Victor Boyarsky, a team member
on the Transarctic expedition and Will's tentmate for more than 500 nights.
According to Will, Victor (a celebrated polar scientist, poet, and author
of five books) is a big personality—big laugh, big hands, big appetite,
big heart—who prefers simple, homey dinners with friends over formal
diplomatic affairs. This borscht recipe is inspired by the classic Russian
soup from Victor's homeland.*

2 tablespoons butter or vegetable oil

1 large onion, diced

1 cup diced celery

1 large leek, diced

Salt and freshly ground black pepper

4 cloves garlic, minced

1 tablespoon tomato paste

1 teaspoon paprika

1 bay leaf

1 thyme sprig or ½ teaspoon dried thyme

1 pound Yukon Gold potatoes (about 6),
   peeled and cut into 1-inch chunks

1 pound medium beets (about 6),
   scrubbed and cut into 1-inch chunks

2 medium carrots, cut into 1-inch chunks

4 to 6 cups Vegetable Stock (page 159) or chicken stock

2 cups chopped red cabbage

1 tablespoon apple cider vinegar, or to taste

Put the butter or vegetable oil into a large, heavy soup pot or Dutch oven and set over medium-high heat. Add the onion, celery, and leek, and stir to coat. Season with salt and pepper and cook until the onion is soft and just starting to brown, about 5 to 7 minutes. Stir in garlic, tomato paste, paprika, bay leaf, and thyme, and cook another minute.

Add potatoes, beets, carrots, and stock and bring to a boil; reduce the heat and simmer until the vegetables are tender, about 20 minutes. Stir in cabbage; taste and adjust the seasoning, then simmer until the cabbage is tender crisp, about 8 minutes. Serve topped with the Horseradish Cream.

## Horseradish Cream

   2 tablespoons prepared horseradish
   1 tablespoon lemon juice
   1 cup whole milk Greek yogurt or sour cream

In a small dish, stir all ingredients together.

# Spaghetti Squash Gratin

Serves 4 to 6

*This hearty casserole gets plenty of crunch and flavor from the "almond breadcrumbs," but feel free to substitute wheat breadcrumbs if you wish. Steaming the spaghetti squash (and any winter squash) whole makes it easier to cut.*

### ALMOND BREADCRUMBS

½ cup almond flour

1 teaspoon Italian seasoning

¼ teaspoon salt

1½ tablespoons water

1 tablespoon olive oil

### FILLING

2 medium spaghetti squash

2 tablespoons olive oil

2 cups marinara sauce

1 teaspoon dried thyme

½ teaspoon red pepper flakes

1 teaspoon salt

¼ teaspoon black pepper

¾ cup shredded Parmesan
  or a mix of Parmesan and mozzarella cheeses

## TO PREPARE THE ALMOND BREADCRUMBS

In a small bowl, combine almond flour, Italian seasoning, and salt. Add the water and mix together until the mixture resembles wet sand. Heat a skillet over medium heat and add the almond mixture. Cook, stirring nonstop for a few minutes until lightly browned. Stir in the oil, then remove from the heat and cool.

## TO PREPARE THE FILLING

Preheat the oven to 425 degrees. Generously grease a 9 × 13-inch casserole pan. Poke a few holes into the squash and set on a baking sheet. Bake until the squash is fork-tender and begins to collapse. Remove and cool to room temperature. Cut the squash in half; remove and discard the core and seeds.

In a small bowl, whisk together olive oil, marinara, thyme, red pepper flakes, salt, and pepper. Using a large spoon, scoop out the strands of spaghetti squash flesh into a large mixing bowl. Add the sauce and gently stir to coat the spaghetti squash strands, then turn into the prepared baking pan. Distribute the cheese over the squash and bake for about 20 minutes. Scatter the almond breadcrumbs over the top and return to the oven to bake until golden brown, about 20 minutes.

# Brussels Sprouts and Tofu Stir-Fry

Serves 6

*Even the most stubborn tofu avoiders will take a shine to this*
*bold stir-fry. Delicious on brown rice, white rice, and rice noodles.*

### STIR-FRY SAUCE

¼ cup mirin

¼ cup soy sauce

3 tablespoons rice vinegar

1 teaspoon toasted sesame oil

### STIR-FRY

2 blocks (14 ounces each) extra firm tofu, drained

2 tablespoons soy sauce

1 tablespoon cornstarch or arrowroot

1 teaspoon red pepper flakes

4 tablespoons vegetable oil

1 pound Brussels sprouts, trimmed and quartered,
   about 4 cups

6 scallions, white parts chopped into
   1-inch diagonal pieces, green parts sliced thin

2 cloves garlic, smashed and rough-chopped

1 tablespoon grated fresh ginger

1 teaspoon sesame seeds

## TO PREPARE THE STIR-FRY SAUCE

In a small bowl, whisk together mirin, soy sauce, vinegar, and toasted sesame oil. Set aside.

## TO PREPARE THE STIR-FRY

Cut the tofu into 1-inch squares. Layer a clean kitchen towel over a large cutting board and put the tofu on the towel; cover tofu with another clean kitchen towel and press down. Allow the towels to absorb the moisture from the tofu for about 15 to 20 minutes; the towels should be soaked through.

In a medium mixing bowl, whisk together soy sauce, cornstarch, and red pepper flakes; add the drained tofu and toss gently to coat. Set aside.

In a large wok or skillet, heat 2 tablespoons of oil over medium-high heat. Add the Brussels sprouts and white part of the scallions and cook, stirring frequently until sprouts are bright green (about 2 minutes). Toss in ginger and garlic and cook 1 more minute. Transfer the vegetables to a bowl.

Heat the remaining oil in the pan, then add the tofu in a single layer and return to the heat. Do not overcrowd; work in batches if necessary. Cook the tofu, undisturbed, until the tofu is browned on the bottom (about 3 minutes), then flip the tofu and continue cooking until browned, another 3 minutes. Return Brussels sprouts to the pan with the tofu and add the stir-fry sauce. Cook, stirring nonstop, until the sauce has reduced and is absorbed by the tofu, about 2 minutes. Stir in sesame seeds. Serve garnished with the reserved sliced green parts of the scallions.

# Mom's Onion, Apple, and Turnip Bake

Serves about 4 to 6

*This recipe was inspired by my Grandmother Steger's* Mom's Recipes Cookbook. *The dish is updated with a drizzle of balsamic vinegar and a sprinkling of shredded Parmesan cheese to make a light entrée or substantial side dish.*

2 large sweet white onions, sliced ½ inch thick
1 large turnip, sliced ½ inch thick
4 tart apples, cored and sliced ½ inch thick
Salt and freshly ground black pepper
2 tablespoons balsamic vinegar
¼ cup butter, cut into pieces
¼ cup grated Parmesan cheese
¼ cup breadcrumbs (optional)

Preheat the oven to 350 degrees. In a baking dish, layer the onions, turnips, and apples, lightly sprinkle with the salt and pepper, then drizzle with the vinegar. Scatter the butter, cheese, and breadcrumbs over all. Cover with aluminum foil and bake until the apples and onions are very tender, about 30 minutes. Remove the foil and continue baking until the cheese is bubbly and the breadcrumbs turn crisp, about 3 to 5 minutes.

# Rita Mae's Tofu Scramble

Serves 4 to 6

*Super easy and quick, this healthy scramble makes a hearty breakfast and a fast lunch. I like to serve this on warmed corn or flour tortillas with salsa, diced avocado, and Breakfast Potatoes (page 41).*

2 blocks (14 ounces each) extra
   firm tofu
¾ teaspoon chili powder
½ teaspoon ground cumin
½ teaspoon ground turmeric
¼ teaspoon sweet paprika
3 tablespoons nutritional yeast

¾ teaspoon salt
⅓ cup water
2 tablespoons vegetable oil
1 small onion, diced
1 red bell pepper, seeded and diced
Chopped cilantro, for garnish
1 small avocado, diced

Remove tofu from the package, drain, and place between two thick towels folded into the shape of the tofu. Place a plate or bowl on top, then add a heavy block or skillet to weight it down. Let the tofu drain for about 15 minutes. Then unwrap the tofu and crumble it into 1-inch pieces, using a fork to lightly mash it, leaving some chunks.

In a small bowl, whisk together spices, nutritional yeast, salt, and water to make a sauce.

Heat the oil in a large skillet over medium heat. Add onion and pepper. Cook, stirring occasionally, until softened, about 3 to 4 minutes, then add the tofu and cook for 2 minutes, stirring frequently. Fold in the sauce and cook until lightly browned, about 7 minutes. Remove from heat and garnish with cilantro. Serve on warm tortillas with a side of Breakfast Potatoes, avocado, and salsa.

I'd always intended to live in the wilderness, but not entirely alone. I knew from the start that I wanted to build a community that appreciated this wildly beautiful and remote place. I'm a teacher at heart, and the mission of the Steger Wilderness Center is to share what I've learned about climate change by providing professional and life skills in sustainability. Rita Mae is integral to our success. Three times a day she dishes up meals for a diverse group—master craftspeople, high school interns, and apprentices from Summit Academy, an inner-city vocational school—as well as visiting poets, storytellers, film crews, journalists, politicians, and interested supporters. For many years, a Japanese environmentalist traveled every summer to work with us. He built his own cabin and planted a Zen meditation garden.

Many of our interns and students encounter the wilderness for the first time here. Some have never before enjoyed such nutritious fresh meals made from scratch, nor have they previously lingered at the table after a meal to talk. One of our students who now returns each summer to teach others told us how The Homestead experience changed his life. He became sober, lost fifty pounds, and now has a career in the trades.

Rita Mae's dishes may be unfamiliar to the crew, but they are comforting, and that's key to a group's happiness and welfare. Sometimes just sharing stories of our favorite comfort foods is a way to connect. On our trans-Antarctic trek in 1989–90, my Chinese tentmate, Qin Dahe, a Nobel Award–winning glaciologist, was always hungry, and he liked to talk about

39

the ingredients in his pantry at home that he missed so much—vinegars, spices, dried and fermented foods. As a young man, he had worked in the rice paddies during the Cultural Revolution. Because staples were scarce, he learned to treat rice with the utmost respect. Rice was sacred, he would remind me, and he insisted I cook it "the right way," never peeking in the pot before the rice was cooked and had "rested." One night I was making our dinner; we were exhausted, and I thought Dahe was napping. Suddenly, just as I was about to lift the lid of the rice pot to take a peek, his fist shot out of his sleeping bag to whack my hand. "No!" he shouted.

# Breakfast Potatoes
# (Not Just for Breakfast)

Serves 4

*The crew knows it's Sunday when the scent of sizzling bacon
and coffee draws them to the Lodge for a relaxed brunch.*

*While many recipes suggest steaming the potatoes in advance, you
can save time and fuss by cooking them in a covered pan, so they steam
and become tender, then removing the lid to fry them until browned.*

4 large potatoes—waxy potatoes
   such as Yellow Fin, Yukon Gold, Red
¼ cup bacon fat, lard, duck fat, or vegetable oil
1 tablespoon finely chopped thyme
1 tablespoon finely chopped sage
Salt and freshly ground black pepper to taste

Cut the potatoes into ½-inch cubes. Add 2 tablespoons of the fat to a large,
heavy skillet and set over medium-high heat. Add the potatoes and turn
to coat with the fat. Cover and cook until the potatoes are just tender,
about 10 minutes, stirring occasionally. If the potatoes begin to stick, add
a tablespoon of water. Remove the lid, increase the heat slightly, and add
the remaining fat. Sprinkle the chopped herbs over the potatoes and cook,
stirring until the potatoes are browned on all sides. They should be very
crisp and cooked through. Serve piping hot, seasoning with salt and pepper
to taste.

# Red Flannel Hash

Serves 2 (and is easily expanded)

*Inspired by cozy flannel work shirts, this hash is a mix of potatoes, sweet potatoes, and golden beets. Other roots, such as carrots, parsnips, and celeriac, also work well. Top these off with a fried egg that will burst into a lush velvety sauce.*

    2 tablespoons olive or vegetable oil
    1 large onion, chopped
    1 medium Yukon Gold potato, scrubbed, cut into ½-inch dice
    1 medium sweet potato, scrubbed, cut into ½-inch dice
    2 medium golden beets, cut into ½-inch dice
    Salt and freshly ground black pepper to taste
    Pinch red pepper flakes, to taste
    Salsa or hot sauce for serving (optional)

In a deep heavy skillet set over medium heat, add the oil and then the onions and cook until transparent, about 3 minutes. Stir in the diced vegetables, and sprinkle with salt and pepper and a pinch of the red pepper flakes. Shake the pan to distribute the vegetables evenly. Cover the pan and reduce the heat, cooking until the vegetables have become tender, about 5 to 10 minutes. Remove the cover, stir the vegetables, and continue cooking until they become slightly browned on all sides. Serve with a side of salsa or hot sauce and top with a fried egg if you like.

# Avocado Toast with Fried Eggs and Blistered Cherry Tomatoes

Serves 1

*"The Rita Mae Breakfast" is a favorite any time of day: it's pretty and it's quick.*

| | |
|---|---|
| ½ avocado | 7 whole cherry tomatoes |
| Generous pinch salt | 1 piece sourdough bread |
| 2 tablespoons olive oil, divided | 1 teaspoon balsamic vinegar |
| 2 eggs | Salt and pepper, to taste |

Mash the avocado with a pinch of salt on a small plate and set aside.

Add 1 tablespoon of the oil to a medium skillet and set over medium-high heat. Crack in one egg at a time and cook until the whites are set and no longer jiggle when you shake the pan. With a large spatula, gently flip the eggs. Turn off the heat and allow the eggs to finish cooking, about 15 to 30 seconds. Remove eggs from pan to a plate.

Toast the sourdough bread.

Set the same skillet over medium-high heat and film it with the remaining oil. Add the tomatoes and cook, undisturbed, shaking the pan occasionally until the tomatoes are golden brown and begin to pop, about 2 to 3 minutes.

Turn off the heat. Add the balsamic vinegar and cook for another 10 seconds while shaking the pan. Season with salt and pepper and remove from the pan to a plate. Spread the mashed avocado onto the sourdough toast. Season with salt and pepper. Serve with the fried eggs and blistered tomatoes on the side.

# Tomato Tofu with Scallions

Serves 4

*This is a Vietnamese dish you won't find on the menus of city restaurants. It's served in homes throughout the country's rural villages in the northern mountains. The sweetness of the onions helps to balance the acidity of the tomatoes. Serve on a bed of jasmine rice with soy sauce for seasoning.*

| | |
|---|---|
| 1 block (16 ounces) extra firm tofu | ½ teaspoon red pepper flakes |
| 5 scallions | 3 tablespoons soy |
| 1 small yellow onion | or 2 teaspoons fish sauce |
| 2 cloves garlic | ½ cup water |
| 8 tomatoes | 1 teaspoon salt |
| 2 tablespoons vegetable oil | |

Cut the tofu lengthwise into ½-inch-thick slices. Cut each slice into 4 quarters. Cut the squares in half diagonally to turn them into triangles. Set aside.

Cut the white parts of the scallions into 1-inch pieces, reserving the green parts and chopping for garnish. Thinly slice the yellow onion lengthwise. Smash the garlic, remove the skins, and mince. Cut each tomato into 8 bite-sized pieces, turn into a bowl, and lightly mash to release the juices.

Film a deep skillet with the oil and set over medium-high heat. Add the onions and cook, stirring frequently until soft, about 7 to 10 minutes. Add the garlic and red pepper flakes and cook for 1 minute. Add the tomatoes with their juices, soy or fish sauce, water, and salt, then add the tofu. There should be enough juice to barely cover the tofu; if not, add a little more water. Bring to a boil, reduce the heat, and simmer uncovered so that the liquid is absorbed by the tofu, about 25 minutes. Remove from the heat and serve the tofu and juices over white rice, garnished with the reserved green parts of the scallions.

# Kim Chi's Fried Rice

Serves 4 to 6

*When I was growing up, my mom, Kim Chi, would make me stir-fried rice for breakfast before I went to school. It's a great way to enjoy leftover rice and the odds and ends of fresh vegetables.*

3 tablespoons vegetable oil

4 cloves garlic, smashed

1 cup chopped fresh vegetables
   (carrots, peas, mushrooms,
   broccoli)(optional)

¾ cup leftover cooked white rice,
   cold

2 eggs, beaten

3 tablespoons soy sauce

½ teaspoon red pepper flakes

1 tablespoon dark sesame oil

Salt and freshly ground black
   pepper, to taste

½ cup minced cilantro

Set a large wok skillet over high heat, add the oil, and when it begins to shimmer toss in the garlic and cook, stirring, until just golden, about 1 minute. Add the vegetables (if desired) and toss, cooking until just coated with the oil, then add the rice, breaking up clumps with a spoon as you go, and fry until it's hot and well coated with the oil, about 3 minutes. Make a well in the center and break the eggs into it. Without disturbing the rice, cook the eggs, stirring them nonstop until they are halfway cooked. Fold the eggs into the rice and cook for a couple more minutes, continuing to stir. Add the soy sauce, red pepper flakes, dark sesame oil, and salt and black pepper to taste. Continue cooking, stirring frequently for another 5 minutes. Serve garnished with the cilantro and extra soy sauce.

•  •  •

**TO COOK RICE** Turn 1 cup long grain white rice into a fine mesh strainer and run under cold water until the water runs clear. Turn the rice plus ½ teaspoon salt and 1¼ cups water into a large pot and set over high heat. Bring the water to a boil, cover with a tight-fitting lid, reduce the heat to low, and cook the rice, undisturbed (and that means not opening the lid), for 18 minutes. Remove the pan from the heat and let the rice stand for at least 15 minutes or until ready to use. Uncover and fluff the cooked rice with a fork before serving.

# Frittata with Warm Spices

Serves 4 to 6 (and is easily expanded)

*A frittata is much easier to make than an omelet or a quiche. There's no tricky timing, no crust to fuss with. You can vary the ingredients with whatever you have on hand. Delicious hot or at room temperature, frittatas are a mainstay for breakfast and lunch at The Homestead. This recipe is flexible—swap out the cheeses and seasonings as you please.*

6 eggs

1½ teaspoons coarse sea salt

1 teaspoon freshly ground
   black pepper

½ teaspoon cinnamon

¼ teaspoon ground nutmeg

½ teaspoon ground cardamom

½ teaspoon ground turmeric

½ teaspoon ground cumin

Generous pinch red pepper flakes

½ cup finely chopped parsley

½ cup finely chopped cilantro

1 cup chopped fresh spinach

¼ cup chopped scallions
   (white part only)

¼ cup finely diced chives

2 to 4 tablespoons extra virgin
   olive oil or butter

1 medium onion, finely chopped

Greek yogurt or crumbled feta
   for serving

Preheat the oven to 400 degrees. In a large bowl, whisk together eggs, salt, black pepper, cinnamon, nutmeg, cardamom, turmeric, cumin, and red pepper flakes; stir in parsley, cilantro, spinach, scallions, and chives. Set aside.

Heat enough oil to generously cover a 10- to 12-inch cast-iron or oven-safe nonstick skillet over medium heat. Cook the onion until just soft, about 2 minutes.

Pour the egg mixture over the onions; using a spatula, spread the mixture over them. Cook over medium heat until the bottom becomes firm and set, about 3 to 5 minutes. Slide into the oven and bake until the top is set, about 3 to 5 minutes.

# Rita Mae's Bacon and Caramelized Onion Quiche

Makes one 9-inch quiche

*This creamy, bacon-studded quiche is baked in a rich yet delicate (and gluten-free) almond-flour crust flecked with rosemary and red pepper flakes.*

### CRUST

3 cups almond flour

¾ teaspoon salt

2 teaspoons dried rosemary

¼ teaspoon crushed red pepper flakes

3 tablespoons olive oil

¼ cup water, or more as needed

### FILLING

2 tablespoons unsalted butter

2 large onions, diced

Salt

12 ounces bacon, cut into ½-inch pieces

10 eggs

⅓ cup milk

1 tablespoon heavy cream

½ teaspoon salt

Preheat the oven to 350 degrees. Lightly grease a 9-inch pie plate.

## TO MAKE THE CRUST

In a large bowl, stir together almond flour, salt, rosemary, and red pepper flakes. Stir in the oil and then the water to form a stiff dough. Turn the dough into the pie plate and press the dough evenly over the bottom and up the sides; a measuring cup works well for this. Bake the crust until it is firm but not yet golden (about 20 minutes), rotating about halfway through. Remove and set aside.

## TO MAKE THE FILLING

In a large pan set over medium heat, melt the butter and add half of the diced onions; cook until they begin to soften, about 2 minutes. Add the remaining onions and a few pinches of salt. Lower the heat to medium low and cook, stirring every few minutes, until the onions turn dark brown, about 15 to 20 minutes. Remove from the heat and allow to cool.

Set a large pan over medium heat and cook the bacon until browned and crisp, about 7 to 10 minutes. Remove the pan, drain the bacon on paper towels, and reserve.

Lower the oven to 325 degrees. In a medium bowl, whisk together the eggs, milk, heavy cream, and salt. Fold in the onions and bacon, and pour the mixture into the crust. Bake the quiche until the center is set and the top is golden, about 1 hour. Allow to cool for 10 minutes before serving.

# Spanakopita Pasta

Serves 4 to 6

*Adding chopped greens gives this pasta dish a fresh lift.*
*You can make it ahead, then finish it off last minute in*
*the oven to give it a lovely toasty brown crust.*

3 tablespoons unsalted butter

8 cups chopped greens
   (spinach, Swiss chard, kale, etc.)

1 cup chopped parsley

Salt and freshly ground black pepper

1 pound fusilli or any corkscrew pasta

8 scallions, trimmed, white and dark
   green parts thinly sliced

4 cloves garlic, thinly sliced

8 ounces cream cheese,
   cut into ½-inch pieces

1 tablespoon fresh lemon juice

4 ounces mozzarella, grated

4 ounces crumbled feta

¼ cup chopped kalamata olives

Bring a large pot of salted water to a boil. Heat the oven to 450 degrees.

Generously butter a 9 × 13-inch baking dish. In a medium bowl, toss together the greens and parsley with a little salt and pepper.

Cook the pasta in the boiling water until just al dente; reserve 1 cup of the pasta water. Drain the pasta and set aside. Return the pot to the stove.

Over medium heat melt the remaining butter in the pot. Then add the scallions, garlic, and a pinch of salt and sauté until softened, about 4 to 5 minutes. Add cream cheese, reserved pasta water, and lemon juice and stir until smooth. Stir in the greens, half the mozzarella, half of the feta, and the chopped olives. Stir in the pasta until everything is combined. Taste and adjust the seasonings.

Transfer the pasta to the baking dish, then top with the remaining mozzarella and feta. Bake until the sauce is thick and bubbly and the top is browned, about 10 to 15 minutes.

# Mom's Favorite Hotdish

Serves 4 to 6

*Will's mom's recipes are simple, straightforward, and plentiful.*
*This is adapted from her spiral-bound* Mom's Recipes Cookbook,
*which Will keeps by his stove. Creamette elbow macaroni is a good*
*choice because it cooks so quickly and holds the sauce.*

1 tablespoon vegetable oil
½ pound lean ground beef
½ pound bulk sausage
¼ cup chopped celery
¼ cup chopped fresh parsley
2 cups chopped fresh tomatoes,
   or 2 cups canned tomatoes, drained and chopped
2 cups cooked elbow macaroni
Salt and freshly ground black pepper, to taste

Preheat the oven to 350 degrees. In a Dutch oven or flame-proof baking dish, heat the oil over medium and add the beef and sausage, breaking it up with a spatula and turning it until it's browned and cooked through. Stir in celery, parsley, and tomatoes, then the pasta. Season to taste with salt and pepper. Bake in the oven until heated through, about 10 to 15 minutes.

# Cabbage, Rutabaga, and Sausage Bake

Serves 6

*Here's the perfect fall dinner and an answer to what to do with the bounty of garden cabbage and rutabagas. Serve this over hot buttered noodles.*

2 pounds Polish sausage, cut into bite-sized slices
1 large green cabbage, cored and cut into 2-inch chunks
1 pound rutabagas, cut into ½-inch rounds
2 tablespoons butter, cut into small chunks
Salt and freshly ground black pepper, to taste
Stone-ground mustard for serving

Preheat the oven to 300 degrees. Arrange the sausage, cabbage, and rutabagas in a 9 × 13-inch baking dish or casserole. Dot with butter and cover with aluminum foil or pot lid. Bake until the cabbage is tender and the sausage is cooked through, about 1½ hours. Remove the cover and continue baking until the sausage is lightly browned and the cabbage is slightly crisped, about 30 more minutes. Serve with the stone-ground mustard, ladled over hot buttered noodles.

# Mom's Meatloaf

Serves 4 to 6

*When my mom immigrated to America from Vietnam, she started to learn classic recipes from my Grandma Steger. This meatloaf recipe is one she made frequently when I was young. She always made extra for the meatloaf sandwiches (page 61) we'd eat the rest of the week, another tradition my grandma passed along.*

Butter or oil for greasing the loaf pan
½ cup grated carrots
1 small onion, grated
2 pounds ground beef
1 cup breadcrumbs
1½ teaspoon salt
1 teaspoon ground black pepper
1 egg
½ cup milk

Preheat the oven to 350 degrees. Lightly grease a loaf pan. In a large bowl combine the carrots, onion, meat, breadcrumbs, salt and pepper, egg, and milk.

Turn the meat mixture into the loaf pan and bake until the inside is no longer pink, about 1 hour and 25 minutes.

# The Rita Mae Reuben

Makes 4 sandwiches

*A messy, two-fisted sandwich, just right for lunch on one of those rainy cold days. Be sure to use a good dark rye bread.*

8 slices rye bread
¼ cup Russian Dressing (page 154)
8 slices Swiss cheese
1 ⅓ cup sauerkraut
1 pound thinly sliced corned beef
¼ cup unsalted butter, softened
4 pickles for serving

Spread 1 slice of bread with ½ tablespoon of the Russian Dressing. Add 1 slice of cheese, ⅓ cup sauerkraut, ¼ pound corned beef, and one more slice of cheese. Spread the second slice of bread with another ½ tablespoon of the dressing, and place dressing side down on the cheese. Spread half the amount of butter on the top slices of bread; flip the sandwiches and place, buttered side down, onto a large heavy cast-iron skillet or griddle. Set the skillet over medium-low heat and cook slowly until the bread is browned and crisped on one side, pressing down with a large spatula. Spread the remaining butter on the top slices of bread. Flip the sandwiches and continue to cook, pressing down with the spatula until crisp and golden. Flip again to reheat the first side for a minute, then remove from the heat and slice in half or quarters. Great with a pickle.

# Best Ever Turkey Burgers

Makes 4 to 6 burgers

*The secret to these juicy burgers is the addition of avocado and mayonnaise in the mix to season and keep the meat moist. Cook these in a skillet, not on a grill that draws out the juices into the fire. Serve with Russian Dressing (page 154) or Burger Sauce (page 157).*

1 large avocado, halved
2 teaspoons Worcestershire sauce
1 tablespoon hot sauce
3 tablespoons mayonnaise
1 teaspoon garlic powder
1 teaspoon salt
1 pound ground turkey

Vegetable oil for the griddle or skillet
4 buns
Russian Dressing (page 154) or Burger Sauce (page 157) for serving

In a large bowl, mash together half of the avocado, Worcestershire sauce, hot sauce, mayonnaise, garlic powder, and salt until smooth. Work in the ground turkey, then refrigerate for 30 minutes.

Remove the turkey mixture from the refrigerator. Smooth the top of the mixture with a large spoon until the surface is flat and divide into 4 portions. Using your hands, shape the mixture into patties.

Film a griddle or cast-iron skillet with oil and set over medium heat. Scoop each burger from the bowl and set on the pan. Cook until browned on the bottom, about 7 minutes, and then, using a spatula, flip each burger. Press the burgers down with the spatula so that the juices cook back into the burgers and continue cooking until brown. Serve the burgers on the buns with slices of the remaining half avocado and Russian Dressing or Burger Sauce.

# Black Bean Wild Rice Burgers

Makes 4 burgers

*This is the burger of choice for omnivores and vegans at Hobo Village gatherings. They are made with the simplest pantry staples. You can also scramble the bean mix in a skillet for nachos and burritos. Be sure to serve the burgers with Burger Sauce (page 157).*

1 cup walnuts

3 tablespoons vegetable oil

1 onion, diced

2 cloves garlic, minced

⅓ cup breadcrumbs or oats

1½ cups cooked or canned
  black beans, drained

1 cup cooked wild rice

1 teaspoon chili powder

1 teaspoon smoked paprika

1 teaspoon cumin

Salt and pepper, to taste

Scatter the walnuts into a large skillet and set over medium heat. Toast the nuts, stirring frequently until they are lightly browned, about 7 minutes. Remove from the pan and allow to cool.

In the same pan, heat 1 tablespoon of oil over medium heat. Add the onion and garlic and cook until the onion is translucent, about 7 minutes; remove from the heat.

Transfer the walnuts to a food processor and process into a fine meal. Add the breadcrumbs or oats and process again until blended.

Turn the black beans into a large bowl and thoroughly mash with the back of a fork. Stir in the wild rice, and add the onion, garlic, chili powder, paprika, cumin, and salt and pepper. Mix together thoroughly.

Divide the mixture into 4 even pieces. Using your hands, roll the pieces into balls and press down to form burgers, holding the patty in one hand and using the other to smooth the sides.

Heat the remaining oil in a pan set over medium heat. Cook the burgers to brown on both sides, about 3 minutes per side.

• • •

**TO COOK DRIED BLACK BEANS** Put the beans in a pot and add enough water to cover the beans by 4 inches. Soak overnight. Drain the beans in a colander, return to the pot, and add enough water to cover the beans by 4 inches. Set over high heat, bring to a boil, reduce the heat to simmer, cover, and cook the beans until tender, about 20 to 40 minutes. Drain the beans and proceed with the recipe.

# Meatloaf Sandwiches

Serves 4

*Whenever I make Will's favorite meatloaf sandwiches, I think about his story of "hopping" freight trains, which is dangerous and illegal. I imagine that Grandma Steger probably had no idea how her son planned to get to Alaska as a college student one summer—she just wanted to give him good food to take on his long journey. She wrapped eight meatloaf sandwiches in newspaper for his backpack.*

*I like to use a good crusty ciabatta roll that holds the thick filling together with lots of Russian Dressing (page 154).*

4 ciabatta buns

½ cup Russian Dressing (page 154)

4 1-inch slices cold meatloaf (page 56)

1 small red onion, thinly sliced

8 leaves romaine lettuce

To assemble the sandwiches, cut the rolls in half crosswise. Generously slather some of the Russian Dressing over one cut side of each. Set the meatloaf slice on top of the dressing, then the onions, then the lettuce. Spread the remaining halves of the rolls with the remaining dressing and set, dressing side down, on top of the lettuce. Press the sandwiches together with the heel of your hand and slice in half or quarters.

Our annual Ice Ball is a necessity. It's the midwinter gathering of friends and family who arrive at The Homestead to help cut and stock ice in the icehouse. I have never used fossil fuels for refrigeration; instead, I've relied on the ice we cut from Picketts Lake for the root cellar. To cut through the thick layers of ice, I found antique ice saws and tongs once used before the advent of electricity in the barns and sheds of neighboring homesteads. The old-timers willingly parted with their tools, knowing that they were going to be put back to good use. They'd invite me in to spend time at the kitchen table, and while drinking coffee I'd soak up their stories of horse teams pulling sled loads of ice from the lakes to the big icehouses used by the whole community.

Ice cutting is labor intensive and I'm always grateful to Lisa Ringer, who brings her workhorses up from the Twin Cities. Once the ice is cut, the horses help haul the huge blocks up the hill to the icehouse, where the team of stackers packs in the blocks and covers them with dry sawdust from the woodshop. Sawdust provides an effective layer of insulation that dramatically slows the melting. The ice will last from that first week of February well past the following October.

A crew of cooks keeps us all well fed through the day with chili and lasagna and bread. At nightfall, we uncork wine bottles, tap the keg, feast, then strike up the music for singing and dancing well into the night. At first light the wolf pack's howls echo across the lake.

It's gotten warmer through the years. Back in 2003, we had more than twenty-four inches of ice, a record thickness. Recently our ice has been far thinner, averaging about eight inches or so, making our work even harder because we have to work in about ten inches of water.

The Homestead has always attracted a range of visitors interested in this work. In the early years, before I had put in a road, traveling here wasn't easy and very few people were willing to make the long trek. One of my favorite guests was Vieve Gore, cofounder of W. L. Gore & Associates (known for innovative products like Gore-Tex) with her husband, Bill, who came up at least once a year. She was a pioneer, one of the first women to run a billion-dollar company on her own, a self-made person. Her parents were homesteaders in Utah, and one of my favorite photographs is of five-year-old Vieve leaning against a musket next to the door of the family's sod house. She was charming and fearless, as happy at a fancy dinner party as she was hiking The Homestead.

My Uncle Harvey also made the long journey from his home in southern Minnesota every fall until the year he turned 100. He would haul in several pounds of the sausage he ground using a recipe he had learned from his German father. Barrel-chested, strong, sharp, and witty, he had a great booming laugh, and I relished his company. He was a fan of The Homestead and told me he appreciated what it took to make a life here. We used the brown paper the sausages were wrapped in to light the grill. He was so positive, and he believed that the secret to his long, happy life was friendship, beer, and sausages.

# Sausage and Bean Stew

Serves 6 to 8

*This is one of those recipes that tastes even better a day or two after it's been made and the seasonings have had time to marry. You can use any sausage you like, such as hot Italian or Polish.*

2 tablespoons oil

1 pound Italian sausage,
 sliced about 1 inch thick

2 tablespoons tomato paste

½ teaspoon ground cumin

2 to 3 carrots, diced into ½-inch pieces

2 stalks celery, diced into ½-inch pieces

1 onion, finely chopped

3 cloves garlic, smashed

Salt and freshly ground black
 pepper, to taste

3 sprigs fresh thyme

1 to 2 sprigs fresh rosemary

1 bay leaf

1 cup white beans

Freshly ground black pepper,
 to taste

Balsamic vinegar, to taste

Film a large stockpot with the oil and set over medium-high heat. Add the sausage and toss until cooked through, about 5 to 7 minutes. Using a slotted spoon, transfer the sausage to a plate covered with paper towels.

Add the tomato paste, cumin, carrots, celery, onion, and garlic to the pot and cook, stirring until coated, about 2 minutes. Season with salt and pepper to taste, then stir in thyme, rosemary, and bay leaf, then the beans with 6 to 8 cups of water. Turn the heat up and bring the liquid to a boil. Then reduce the heat to low and simmer until the beans are tender, about 1 to 2 hours, adding more water to keep the beans covered.

When the beans are tender, return the sausage to the pot and simmer for 5 minutes. Season to taste with pepper and a splash of balsamic vinegar.

**NOTE** You don't have to soak the beans, but it does help speed up the cooking time. Put them into a bowl and add enough water to cover the beans by 4 inches. Soak overnight and drain before adding to the pot.

# Vieve Gore's Black Bean Chili

Serves 6 to 8

*Vieve Gore liked to cook for Will and crew when she visited. This is a favorite.*

2 tablespoons vegetable oil
1 onion, finely chopped
2 medium carrots, cut into small dice
1 red pepper, cored and diced
2 large cloves garlic, minced
3 tablespoons mild ground chili, or to taste
1 tablespoon ground cumin
1 teaspoon dried oregano
1 can (28 ounces) chopped tomatoes
4 cups cooked turtle or black beans, or canned beans, with broth
2 cups diced winter squash or sweet potatoes, about ¾ pound
Salt and freshly ground black pepper, to taste
Grated cheddar cheese for garnish

Heat the oil in a large Dutch oven or stockpot over medium heat. Stir in onion, carrots, and pepper and cook, stirring often until the vegetables are tender and beginning to turn brown, about 8 minutes. Stir in the garlic and cook for about 30 seconds; add the chili, cumin, and oregano. Add the tomatoes and simmer, stirring often, until the tomatoes have cooked down and the mixture thickens, about 10 minutes.

Stir in the beans with some of their liquid, if the mixture needs thinning. Add the winter squash and bring to a simmer, stirring often so that the chili doesn't stick to the bottom of the pot. It should be thick, but not too sticky. Season to taste with salt and pepper. Served topped with cheese.

• • •

**TO COOK DRIED BEANS** Turn 2 cups of beans into a colander and rinse under cold running water. Pick over the beans for small rocks or debris. Turn into a pot and add enough water to cover the top of the beans by 4 inches and soak for 6 hours or overnight. Drain the beans, turn into a pot, and add enough water to cover the beans by 1 inch. Add 1 small onion, quartered, and a bay leaf and bring to a boil. Reduce the heat to a bare simmer, partly cover the pot with a lid, and cook until the beans are tender, about 1 hour, stirring occasionally.

# Homer's Smoked Fish and Pasta Salad

Serves 4 to 6

*This is one of Will's favorite lunch salads, and it can be doubled or tripled for friends and visitors. It's named for the homestead cat who roams the snowy woods in the winters before retreating to the lodge for warmth; summers he lounges outside the lodge, waiting for kitchen scraps, like the smoked fish for this salad. I like to use the firm, flaky Lake Superior whitefish, smoked in Duluth and sold at Zup's grocery store in Ely. The salad comes together quickly.*

6 tablespoons olive oil
Juice of 1 large lemon
1 clove garlic, smashed
3 scallions, finely chopped
1½ cups cherry tomatoes, halved
½ cup finely chopped parsley,
   plus more for garnish

Generous pinch red pepper flakes
9 ounces bow-tie pasta, cooked
   and drained, room temperature
1 pound smoked whitefish,
   skinned, boned, and flaked
¼ cup shredded Parmesan
   cheese

In a large bowl, whisk together oil, lemon juice, and garlic. Toss in scallions, tomatoes, parsley, and a generous pinch of red pepper flakes, then toss in the pasta to thoroughly coat. Toss in the smoked whitefish and Parmesan and serve garnished with remaining parsley.

# Vietnamese Steak Salad with Hard-Boiled Eggs

Serves 4 to 6

*This is the kind of salad you'll find on the menus of food trucks and carts throughout Saigon. The stir-fry sauce becomes a vinaigrette that pulls all the flavors together when tossed with the hard-boiled eggs. For a complete meal, serve this salad with a bowl of white rice.*

2 tablespoons plus
   1 teaspoon soy sauce
1 teaspoon fish sauce
¼ teaspoon sugar
½ teaspoon salt
¼ teaspoon black pepper
1 pound beef steak, thinly sliced
1 head red or green leaf lettuce,
   leaves rinsed, dried

3 medium tomatoes,
   cut into wedges
Juice of half a lemon
2 tablespoons cooking oil
½ large onion, sliced
3 hard-boiled eggs, quartered
A few grinds of black pepper

In a bowl, whisk 2 tablespoons soy sauce, fish sauce, sugar, salt, and pepper together. Marinate the sliced steak in the sauce for 10 minutes; set aside.

In a large bowl, toss the lettuce and tomatoes with lemon juice and 1 teaspoon of soy sauce. Arrange on a serving platter.

In a large skillet set over medium-high heat, add the oil and sauté the onion until tender, about 3 minutes. Add the steak and marinade and sauté, stirring frequently until the steak is cooked through, about 3 to 5 minutes.

Transfer the beef and hard-boiled eggs to the platter with the lettuce and tomatoes and drizzle with the pan juices. Toss the salad until the ingredients are well coated and the egg yolks have begun to break down into the vinaigrette. Season with a few grinds of black pepper.

# Homestead Chickpea Curry

Serves 4 to 6

*Perfect for vegans and vegetarians, this recipe is made from pantry ingredients and pairs well with a side of basmati rice, plain yogurt, and naan. Toss in extra cauliflower, diced potatoes, or any other vegetables of your choice.*

2 tablespoons vegetable oil or ghee

1 large yellow onion, finely diced

2 tablespoons curry powder

3 cloves garlic, minced

1 inch fresh ginger, minced

1 can (13.5 ounces) coconut milk

1 teaspoon salt

1 tablespoon honey

1 red bell pepper, seeded
    and cut into 1½-inch pieces

3 cups cooked dried chickpeas,
    drained, or 2 cans
    (15.5 ounces each)

1 cup green beans,
    cut into 1½-inch pieces

Juice of half a lime

Salt and freshly ground black
    pepper

1 cup unsalted, roasted peanuts

¼ cup chopped fresh cilantro

Heat the oil or ghee in a deep pot over medium-low heat and add onions. Cook, stirring frequently until translucent, about 5 to 7 minutes. Stir in curry powder, garlic, and ginger and cook, stirring frequently, until fragrant, about 2 minutes.

Stir in the coconut milk, scraping up any of the bits sticking to the bottom of the pan. Add the salt and honey and stir well. Increase the heat to medium high, bring to a boil, then reduce the heat and simmer to reduce the liquid, about 15 minutes. Stir in red pepper, chickpeas, and green beans and continue cooking until the vegetables are tender, another 5 to 10 minutes. Season to taste with lime juice, salt, and freshly ground black pepper. Serve over rice garnished with the peanuts and cilantro.

. . .

**TO COOK DRIED CHICKPEAS**  Put the chickpeas into a pot and add enough water to cover by 4 inches to soak the chickpeas overnight. Drain in a colander, turn the chickpeas into a large pot, and add enough water to cover by 4 inches. Set the pot over high heat and bring to a boil; reduce the heat to simmer the liquid. Cover and cook the chickpeas until tender, about 45 minutes to an hour, checking to add more water if necessary. Drain the chickpeas and proceed with the recipe.

# Ice Ball Veggie Lasagna

Serves about 6

*This dish is guaranteed to convert people who swear they "don't like vegetables." It can be made ahead and easily doubled or tripled to feed a crowd.*

1 ½ to 2 pounds fall vegetables
   (use a mix of Brussels sprouts,
   carrots, cauliflower, broccoli,
   turnips, kohlrabi, beets, etc.),
   cut into ¼-inch pieces
3 to 4 tablespoons extra virgin
   olive oil
Coarse salt
8 ounces ricotta cheese

1 egg
Pinch ground nutmeg
Salt and freshly ground black
   pepper
2½ cups marinara sauce,
   or more as needed
8 ounces no-boil lasagna noodles
4 ounces (1 cup) grated Parmesan
   cheese

Preheat the oven to 425 degrees. Line 1 or 2 sheet pans with parchment paper. Toss the vegetables with about 2 tablespoons of oil, enough to evenly coat them. Spread the vegetables on the baking sheets so they don't touch and sprinkle with a couple of pinches of coarse salt. Roast, stirring occasionally, until the vegetables are crisp and brown, about 15 to 20 minutes. Remove and reduce the oven to 350 degrees.

Lightly oil a rectangular baking dish.

In a small bowl, blend together ricotta, egg, nutmeg, and a little salt and pepper.

Spread a tablespoon of marinara sauce over the bottom of the baking dish. Top with a layer of the noodles. Top with a few tablespoons of the ricotta mixture, spreading it over the noodles into a thin layer with a rubber spatula. Top the ricotta mixture with half of the roasted vegetables, then top those with more marinara sauce and sprinkle with Parmesan cheese. Repeat these layers starting with the noodles, then add a final layer of noodles, marinara sauce, and finally the Parmesan. Drizzle a little oil over the top of all.

Cover the baking dish tightly with foil, place in the oven, and bake until the noodles are tender and the mixture is bubbling, about 40 minutes. Remove foil, return to the oven, and bake until the cheese is lightly browned, about 3 to 5 minutes. Remove and allow to rest for about 5 minutes before serving.

# Mom's Mac and Cheese

Serves 4 to 6

*Inspired by the recipes in* Mom's Recipes Cookbook, *this classic favorite uses cream cheese as a shortcut to make a cream sauce with butter and flour. A generous dash of your favorite hot sauce will give this a spicy kick.*

1 pound elbow macaroni
Kosher salt
2 cups whole milk
4 ounces cream cheese, cut into 1-inch pieces
12 ounces sharp cheddar cheese
3 tablespoons butter
Freshly ground black pepper
Dash of hot sauce (optional)

Cook the pasta in a large pot of salted water until just tender and slightly undercooked.

In a large pot, bring the milk up to a simmer and then whisk in the cream cheese until it's blended. Add the cheddar cheese and butter, whisking until completely melted. Season with salt and a generous amount of pepper.

Stir in the cooked pasta. Continue to cook over low heat until the sauce has thickened and coats the pasta, about 3 minutes. The sauce will thicken as it cools. Season with a dash of hot sauce and salt and pepper before serving.

# Indian Butter Chicken

Serves 6

*I always have a blast making this dish because the crew really loves it. The chicken marinates in yogurt and spices for a few hours to make it tender and juicy. The sauce simmers for one hour and is pureed until smooth. As it simmers, the aroma draws workers to the Lodge. It's great with basmati rice.*

### MARINADE

1 cup plain yogurt
1 tablespoon lemon
   juice
1 teaspoon turmeric
2 teaspoons garam
   masala
1½ tablespoons
   ground ginger
4 cloves garlic,
   smashed
Generous pinch salt
   and pepper
2½ pounds bone-in,
   skin-on chicken
   parts
2 tablespoons butter

### SAUCE

½ cup butter (1 stick)
1 cinnamon stick
6 cardamom pods,
   smashed
1 whole clove
2 teaspoons
   fenugreek
2 onions, diced
1 to 2 small, spicy
   peppers,
   seeded and diced
1 tablespoon grated
   fresh ginger
4 cloves garlic,
   smashed

1½ teaspoons garam
   masala
1½ teaspoons ground
   cumin
1 teaspoon paprika
½ teaspoon turmeric
½ cup chicken stock
1 can (28 ounces)
   crushed tomatoes
½ cup heavy cream
   or coconut cream
Chopped cilantro,
   for garnish

In a large bowl, whisk together yogurt, lemon juice, turmeric, garam masala, ginger, and garlic. Season with salt and pepper. Add the chicken and coat with the marinade. Cover and refrigerate for at least 3 hours.

Remove the chicken from the marinade and wipe off the marinade with a paper towel. Discard the marinade. In a large pan set over medium heat, melt the butter until it begins to foam. Brown the chicken on both sides, about 3 minutes per side. Remove the chicken and set aside.

In the same pan, melt the butter for the sauce and add the cinnamon, cardamom, clove, and fenugreek. Cook until fragrant, about 2 minutes. Add onion and peppers to the pan and cook, stirring often until the onions are soft, about 7 minutes. Add the ginger, garlic, garam masala, cumin, paprika, and turmeric, and cook for another 2 minutes. Add the stock and tomatoes and bring to a boil. Lower the heat to simmer, uncovered, until thickened, about 45 minutes to 1 hour. Remove the sauce from the heat and allow to cool for a few minutes. Working in batches, puree the sauce until smooth. Return the sauce to the pan, bring to a simmer over low heat, and stir in the cream. Add chicken to the sauce and continue simmering until the chicken is cooked through, about 15 to 20 minutes. Serve over white basmati rice and garnish with cilantro.

**GARAM MASALA,** like curry powder, is a spice mix that adds flavor and color to a dish. It is a blend of coriander, cumin, black pepper, whole cloves, cardamom, and fennel. It is available in the spice aisle of most grocery stores.

# Sausage, Onion, and Apple Scramble

Serves 4 to 6

*Use any sausage you like for this simple fall dish inspired by Will's stories of Uncle Harvey.*

¼ cup butter

2 large yellow onions,
    cut into ¼-inch slices

Salt and freshly ground black
    pepper

2 sprigs fresh thyme

3 sprigs fresh parsley

5 large, tart apples
    (e.g., Haralson or Cortland),
    cored and cut into ½-inch slices

8 bratwurst or Italian sausages,
    about 4 ounces each

½ cup beer, white wine, or water

Preheat the oven to 250 degrees. Put 2 tablespoons of butter into a large cast-iron pan or heavy skillet and set over medium-high heat. Add the onions and season with salt and pepper; stir to coat with butter. Add the thyme and parsley; reduce the heat and cook, stirring frequently until the onions are nicely browned, about 10 minutes. Remove from the pan, set in a baking dish, and hold in a warm oven.

Put the remaining butter into the pan and add the sliced apples. Increase the heat to medium-high and cook, stirring the apples, until they are brown on both sides, about 6 to 8 minutes. Remove from the pan, add to the baking dish with the onions, and hold in a warm oven.

Prick each sausage in several places with the tip of a sharp knife. Return the pan to the stove over medium heat and add the sausages in one layer. Brown the sausages slowly on all sides, about 8 to 10 minutes. Add the beer, wine, or water and cook until reduced to a glaze.

To serve, arrange the apples and onions on a large platter and place the sausages over the onion–apple mixture.

# Ely Porketta

Serves 8 to 12

*We buy preseasoned porketta roast at Zup's Food Market in Ely. It reflects the Italian heritage of the miners who settled the small towns along the Iron Range, planted huge gardens, and raised hogs and goats.*

*To make this from scratch, season the meat the night before, wrap it tightly in plastic, and store in the refrigerator. Be sure to bring it to room temperature before roasting. It will emerge with golden cracklings. Plan on leftovers—great for sandwiches and tacos.*

1 7- to 8-pound bone-in, skin-on pork shoulder roast, fat trimmed to ¼-inch thickness
5 cloves garlic, cut into slivers
Finely grated zest of 1 lemon
2 teaspoons crushed fennel seed
Generous pinch red pepper flakes
Coarse salt and freshly ground black pepper
6 sprigs fresh rosemary
6 sprigs fresh thyme

Score the skin and fat all over the pork, taking care not to cut down into the meat. Using a sharp knife, cut enough slits into the pork to fill with garlic slivers. Rub the lemon zest, crushed fennel seed, red pepper flakes, salt, and pepper over the pork. Press the rosemary and thyme into the pork. Wrap with plastic, set on a large plate, and refrigerate overnight.

Remove the pork from the refrigerator, transfer to a rimmed baking sheet, and allow it to come to room temperature. Preheat the oven to 450 degrees.

Roast the pork for 35 minutes, reduce the temperature to 325 degrees, and continue cooking until a meat thermometer inserted into the thickest part of the meat reads 180 degrees, about 3 to 4 hours.

Transfer the meat to a cutting board and serve with the cracklings.

# Very American Goulash

Serves 4 to 6

*My mom, Kim Chi, has a knack for spicing up American recipes, adding a dash of soy sauce and a sprinkle of red pepper flakes to boost flavors. Serve this over buttered noodles.*

1 tablespoon vegetable oil
1 pound ground beef
1 onion, diced
1 can (28 ounces) whole tomatoes
¼ cup tomato paste
2 tablespoons soy sauce
1 teaspoon red pepper flakes
1 teaspoon salt
4 cloves garlic, smashed

Film a large deep skillet with the oil and set over medium heat. Crumble the ground beef into the skillet and cook until it is no longer pink. Stir in onion and cook until softened, then add tomatoes, tomato paste, soy sauce, red pepper flakes, salt, and garlic. Lower the heat and simmer until the liquid is slightly reduced, about 10 minutes. Taste and adjust the seasonings.

# Beef One Pot with Asian Spices

Serves 4

*This stovetop casserole, seasoned with Asian flavors—ginger, soy, and mirin—comes together quickly and is perfect on a winter night when the temperatures plummet below zero. Serve over wild rice or noodles.*

 1 to 2 tablespoons vegetable oil
 2 pounds boneless beef chuck, cut into 1-inch chunks
 2 cups beef or chicken stock
 ¼ cup soy sauce
 ¼ cup mirin
 10 slices of fresh ginger
 A few grinds of black pepper, to taste
 1½ pounds peeled butternut squash or sweet potatoes
 Juice of a small lemon
 Salt and freshly ground black pepper, to taste

Film a large skillet with the oil and set over medium-high heat. Sear the meat until nicely browned on all sides, about 5 to 8 minutes. Transfer the chunks to a medium-sized casserole.

Add the stock to the skillet, increase the heat, and stir, scraping all the solids into the liquid. Add this to the casserole with soy sauce, mirin, ginger, and a few grinds of black pepper. Cover and simmer on top of the stove, stirring occasionally, until the meat is tender, about 45 minutes. Then stir in the squash and continue cooking until tender, about 15 to 20 minutes. Season to taste with the lemon juice, salt, and pepper.

# Cashew Beef Stir-Fry

Serves 4 to 6

*Here's a streamlined and flavorful recipe for a favorite takeout dish: wok-fried beef in a potent sauce. Just right for a cool fall night in the north woods.*

½ cup unsalted raw cashews

6 tablespoons hoisin sauce

3 tablespoons soy sauce

1 teaspoon rice vinegar

2 tablespoons tomato paste

1 teaspoon toasted sesame
　　seed oil

1 tablespoon vegetable oil

1 onion, diced

2 pounds ground beef

1 teaspoon red pepper flakes

4 cloves garlic, minced

1 tablespoon freshly grated
　　fresh ginger

6 scallions, white parts only,
　　chopped; green parts chopped
　　for garnish

1 tablespoon sesame seeds

Preheat the oven to 350 degrees. Scatter the cashews over a baking sheet and roast, shaking the pan frequently, until golden and fragrant, about 10 to 15 minutes. Set aside.

In a small bowl, whisk together hoisin sauce, soy sauce, vinegar, tomato paste, and toasted sesame seed oil.

Heat the vegetable oil in a large skillet set over medium-high heat and cook the onion until translucent, about 7 minutes. Add the beef and cook, breaking it up as you go, stirring frequently until browned, about 5 minutes. Toss in red pepper flakes, garlic, ginger, and chopped white scallions, and cook until fragrant, about 1 minute. Stir in the hoisin sauce mixture and cook until it's absorbed by the beef, about 3 minutes. Remove the pan from the heat and toss in the cashews. Serve over jasmine rice, garnished with the chopped scallion greens and sesame seeds.

W hen the Wilderness School was in session, we'd all sleep outside by the fire—students, guides, and dogs. Those clear and still winter nights were especially magical. The puppies would tumble and play before snuggling down while we sat transfixed by the crackling flames sending sparks up into the dark.

When Patti's sons were young, we celebrated Christmas together in the cabin. Christmas afternoon the boys and I would go into town and meet Patti at her sewing shop. We'd stock up on food and then we would all drive back in the old '49 GMC pickup to the landing on the first lake and hike over the second lake and up to the cabin. One year it was so cold and windy, with temperatures below negative 20 degrees, that we had to cover the boys' heads with brown paper bags to protect their faces

from the wind chill. When we were inside we got a roaring fire going in the woodstove, lit candles on the Christmas tree, and opened our homemade gifts (knitted hats and mitts). We feasted on beef roasted in the wood-fired oven and sang Christmas songs. But the best was playing strip poker—the loser (always me) had to run around the cabin outside naked in the snow. I'm not a religious person, but fires are sacred at The Homestead: they draw us together with warmth and light to tell stories, recite poetry, sing. They seem to bring out the best in us when we gather together as night closes in. Bonfires are central to The Homestead community life.

# Christmas Roast Beef

Serves 4 to 6

*At The Homestead, we always use beef from cattle raised on pasture. In fact, we rely on pastured animals for all of our meat, poultry, eggs, and dairy. Pasture's grasses and flowers help keep cover on the land year-round to stem the runoff of topsoil, capture water, and provide a habitat for pollinators. These vigorous plants capture carbon from the atmosphere and send it back into the ground, where it nourishes the soil. All of this helps mitigate the impact of climate change and regenerates our land.*

*When making this roast, be sure to get the oven plenty hot before you start. Serve with Breakfast Potatoes (page 41).*

1 beef roast (top, eye, or bottom round),
    about 2⅓ to 3 pounds
Coarse salt
Freshly ground black pepper
3 cloves garlic, peeled and minced
1 tablespoon hazelnut or vegetable oil
Red pepper flakes, to taste

Bring the roast to room temperature. Preheat the oven to 500 degrees. In a small bowl, stir together salt, pepper, garlic, and oil to create a paste. Rub this all over the roast. Put the beef into a roasting pan or cast-iron skillet, fat side up, and put into the oven. Cook for 5 minutes per pound.

Turn off the oven and leave the roast to continue cooking for 2 hours (undisturbed!). Then remove the roast and slice.

# Firepit Roast Chicken

Serves 4

*Cooking whole chicken over an open fire guarantees the skin will be crisp and the meat juicy, with hints of wood smoke and spice. It's a dramatic yet remarkably easy method as long as the chicken is well seasoned and trussed.*

| | |
|---|---|
| 1 3- to 4-pound chicken | Red pepper flakes |
| 2 to 3 tablespoons extra virgin olive oil | Coarse black pepper |
| | Butcher twine (available at most grocery and cooking stores) |
| Coarse salt | |
| Paprika | |

Rub the chicken all over with oil and rub in the seasonings. Set the chicken in the refrigerator and allow it to sit, uncovered, overnight.

To truss the chicken: cut the string into a 5-foot segment. Soak the string in water. Lay the chicken on a clean surface, setting it on its back on the center of the 5-foot string. Cross the string over the middle of the chicken, around the middle of the legs, and pull hard so that the legs are pushed slightly up. Wrap the string once around the bottom of the legs, starting from the outside and then going around and toward the inside. Try to keep pressure on the string as you go so it doesn't loosen. Bring both ends of the string over the neck and around the tail until coming back to the center point. Make sure the chicken is wrapped tightly.

Set up an outdoor fire, using oak or hickory wood, to reach a temperature of about 400 degrees.

Hang the chicken (using string and an S hook) about 3 to 4 feet above the fire. A tripod will also work. Cook the chicken, turning it with a stick, for about 4 to 5 hours. When cooked, the internal temperature of the chicken will be 165 degrees on a meat thermometer.

# Ember-Roasted Corn

Serves 6

*There is no better way to cook local corn than over the glowing embers. The smoke-kissed kernels turn sweet and caramelized, and any leftovers are perfect in Wild Rice Salad (page 17).*

6 ears fresh corn, in the husk
12 tablespoons unsalted butter
Coarse salt
Freshly ground black pepper

Gently pull the husks away from the corn but do not detach from the cob. Remove the silk. Pull the husks back up over the kernels. Secure the husks tightly with a string or strip of foil and soak in a large pot of room-temperature water for at least 15 minutes and up to an hour.

Build a hot fire and allow it to burn down until the coals are completely covered with ash. Lay a grill over the embers and place the corn in a single layer over the coals. Roast, turning the corn occasionally, until the husks are charred and the kernels seem tender underneath the husks, about 10 to 15 minutes. Remove the corn, peel back the husks, and serve the corn with plenty of butter, salt, and pepper.

# Spatchcock Chicken with Blueberry Maple Glaze

Serves 4 to 6

*This method of butterflying a chicken ("spatchcock") reduces the cooking time for a whole bird by half and ensures it browns evenly because more of the surface is exposed to the heat. This is as easily done in a big cast-iron skillet over the open fire as it is in the oven. The trick is to begin with a blast of heat to sear the exterior, then continue cooking over lower heat. A meat thermometer is helpful; when done, the dark meat should be roasted to 165 degrees. Finish with a sweet-tangy maple glaze.*

3 to 3½-pound chicken
6 tablespoons vegetable oil
4 cloves garlic, smashed
2 tablespoons chopped
   fresh rosemary
2 tablespoons chopped
   fresh thyme

Salt and freshly ground
   black pepper
½ cup white wine
I cup fresh blueberries
2 to 3 tablespoons maple syrup
Parsley for garnish

Place the chicken, breast side down with the neck of the chicken facing you, on a clean work surface. Using kitchen shears, cut along the length of the backbone on both sides and remove. Turn the chicken over, breast side up. Open it on the counter and flatten the chicken as much as you can with your hands.

Film a cast-iron skillet with oil and set over medium-high heat. Place the chicken in the skillet, breast side down, and cook until you hear it sizzle, about 1 to 2 minutes. Reduce the heat to medium low and cook for 15 to 18 minutes. Using tongs, turn the chicken over, breast side up, and cover with garlic and herbs; sprinkle with salt and pepper.

Add the wine to the skillet and cook another 20 minutes, basting with pan juices, then add the blueberries and stir, spooning them over the chicken. The chicken is fully cooked when a meat thermometer inserted into the thigh registers 165 degrees. Stir in maple syrup and baste the chicken several times before removing from heat. Allow the chicken to rest about 5 minutes before serving with berries and juices spooned over all and garnished with parsley.

# Pan-Roasted Fish with Herb Butter

Serves 4

*The Homestead apprentices and crew like to fish Picketts Lake after work. It's a clean, spring-fed northern lake, home to northern pike and yellow perch. These delicate and light whitefish are best cooked in a cast-iron skillet with butter and a few herbs over an open fire. This simple technique works with any fillet, skin on or off, as long as it's not too thick.*

4 fish fillets, about ½ inch thick
Salt and freshly ground black pepper
3 tablespoons vegetable oil
3 tablespoons butter
A few sprigs fresh thyme or parsley
Chopped parsley, for garnish
Lemon wedges

Pat the fillets dry with a paper towel. Season on both sides with salt and pepper. Heat a heavy 10-inch skillet or cast-iron pan over high heat. Add the oil, place the fillets in the pan (skin side down if skin is on), and press gently with a spatula.

Lower the heat to medium and let the fish sizzle until golden around the edges, about 2 to 3 minutes. Carefully flip the fillets and add butter and thyme to the pan. Tilt the pan slightly to let the melted butter pool to one end. Use a spoon to baste the fish with the butter. Continue basting until the fish is golden all over and cooked through, about 45 to 90 seconds more depending on the thickness. Serve garnished with chopped parsley and lemon wedges.

# Fire-Roasted Baba Ghanoush

Serves 6

*This will be chunkier than most baba ghanoush recipes,
but it's fabulous on flatbreads and as a sauce for baked potatoes
and roasted squash.*

1 large eggplant
¼ cup tahini
Juice of 1 large lemon
2 cloves garlic, minced
Olive oil
Salt and freshly ground black pepper, to taste
2 to 3 tablespoons chopped fresh parsley for garnish

Build a hot fire and allow it to burn down until the coals are completely
covered with ash. Wrap the eggplant in aluminum foil and roast on the grill
until very soft when poked with a fork, about 45 to 60 minutes. Remove and
allow to cool, then unwrap the eggplant. Cut off the top, peel, discard the
skin, and put the flesh into a large bowl. Using a fork or potato masher,
mash in tahini, lemon juice, garlic, and enough oil to make a creamy paste.
Season to taste with salt and pepper. Garnish with parsley before serving.

## Johnny R's 'Rooms on Toast

*If I'm lucky enough to arrive in Ely in the spring when the morels are in season, John Ratzloff, The Homestead photographer, will take me hunting. He can spot a herd of morels a mile away. Along with the big cloth sack, he packs a cast-iron skillet, good butter, salt, pepper, baguette, and a bottle of red wine. Once that sack is full, he'll build a small fire and sizzle up the mushrooms to eat straight out of the skillet, using torn baguette to absorb the juices. No recipe needed!*

# Homestead Steak and Morels (Black and Blue)

Serves 2

*A couple of times each week through the summer, Will and John and I cook dinner together in "Hobo Village" on the lakeshore. We share stories and sip wine as the wood burns low, then we cook over the coals. John made this steak dinner one night using the morels he had foraged and dehydrated. We rehydrated the mushrooms, cooked them with the steak, and used their soaking liquid for a pan sauce. The best way to cook steak is in a heavy cast-iron skillet over a wood fire!*

1 steak, New York cut,
   1½ inches thick, about 1½ pounds, room temperature
Generous pinch coarse salt
1 tablespoon whole black peppercorns, cracked
1 tablespoon vegetable oil
3 garlic cloves, smashed
3 tablespoons butter
1 scallion, chopped
1 cup sliced cremini mushrooms
1 cup rehydrated and drained morels
   (soaking liquid reserved)
¼ cup brandy or white wine
1 medium tomato
1 tablespoon chopped parsley
Baguette, to sop up juices

Pat the steak dry. Sprinkle with salt and cracked peppercorns, being sure to press them into the meat on all sides.

Heat the oil in a large cast-iron skillet over high heat. Add the garlic and sizzle for 30 seconds until golden. Remove and set aside.

Lay the steak flat in the pan and cook for 1 minute. Lower the heat to medium high and cook for 3 minutes more. Flip and continue cooking the steak for 4 minutes on the second side. Using tongs, stand the steak against the edge of the pan to sear and brown the fat for a few minutes on each side. Remove to a warm platter.

Pour off the fat and remove and discard any loose peppercorns from the skillet. Return the pan to medium-high heat and add 1 tablespoon of butter and the scallion. Toss in the fresh mushrooms and the rehydrated morels, stir, and cook a few minutes. Stir in the brandy or white wine and the reserved mushroom soaking liquid and cook until the liquid has reduced to a glaze. Stir in the remaining butter and cook 1 more minute. Pour the entire mixture over the steak.

Place the skillet back over medium-high heat. Halve the tomato, place on the skillet cut side down, and sear for a couple of minutes. Place the tomatoes on the platter at each end of the steak. Sprinkle the parsley over the steak and serve with the baguette to soak up the pan juices.

In 2008, I launched my Ice Out Expedition, starting in Saskatchewan, destined for Baker Lake in Canada's Nunavut Territory, planning to travel two hundred miles through the tree lines to the open Barrens. I skied, pulling a custom-built canoe with runners on the bottom loaded with provisions to cross the tundra and then paddle the Barrens' major rivers, which can get wild as they flow into the Arctic Ocean.

Skiing through the forest is especially tricky—the snow is deep, crusty, and slushy, while the Barrens are flat, with hard-packed, stable snow. On this trip, the weather and wind were against me the entire time; the soft snow, felled trees, and frequent blizzards made passage nearly impossible. I couldn't have made it through without the support of Jason Cook.

This young Dene man lives in Black Lake Village with his wife, Bernadette, and their family of five kids. Jason had been out hunting when he came across my tent one afternoon, where I was resting. Curious and concerned, he poked his head in to check on me. He knew bad weather was coming and offered to help me out. We headed off to the cabin where Jason was raised by his grandfather after his parents abandoned him. It was snug, made cozy by a huge woodstove fashioned from a gasoline drum cut in half.

Jason hadn't gone to school, he couldn't read or write, but he was a most pragmatic and innovative woodsman and a fine companion. We spent several nights riding out that big storm; when we used up our candles, Jason made a light with lard and a strip from my washcloth attached to the top of a Spam can. We chopped wood and cooked together. He told me about his grandfather, Leon, and he called me "grandfather" several times, a gesture of great respect. Leon was the only adult in his life who loved him unconditionally. Even though it was Easter Sunday weekend and I had kept insisting Jason return to his family for the holiday, he stayed with me. On our last evening, as temperatures hovered at 40 degrees below zero, we talked until sunrise, then stoked the stove for a leisurely last breakfast of bannock; that staple, a flatbread made simply of flour, lard, water, and salt, was absolutely delicious when we cooked it over the open fire.

We ate a lot of bannock those early years at The Homestead, cooked in our first stove, an old barrel with a hole cut into the side that we named Harvey. We used its top like a griddle and called the flatbread Harvey's head. When an old homesteader who lived nearby gave us his wood-fired stove, Patti was able to bake beautiful loaves for the Wilderness School. She would grind flour from the sacks of wheatberries we stored in the cache and she'd knead the dough by hand. You can imagine how good the bread smelled as it baked; the aroma drifted for miles. The bread came out of the oven dense and fragrant with a perfect, crackly crust.

# Steger's Bannock

Makes a 9- to 10-inch flatbread

*Here is a recipe for bannock inspired by Will's story of Jason Cook.
You might try making this bannock with barley or oat flour or cornmeal.
It's best cooked in a cast-iron skillet over an open fire, so it becomes crusty
and smoke kissed. Serve it hot, dripping with butter.*

1½ cups unbleached all-purpose white flour

1½ cups whole wheat flour

2 tablespoons baking powder

1 teaspoon salt

2 tablespoons bacon fat, lard, or butter,
    plus a little more for greasing the pan

1 to 1½ cups milk or water

In a medium bowl, mix together the flours, baking powder, and salt. Cut in
2 tablespoons of the fat until the mixture resembles coarse breadcrumbs.

Set a 9-inch cast-iron skillet over medium heat and add a little fat; it should
sizzle but not smoke. Stir the milk or water into the dry ingredients to make
a thick batter. Turn the batter into the skillet and press it out evenly to be
about 1 inch thick. Cook until the bottom is a dark golden brown and the top
is starting to become dry, about 10 to 15 minutes. Flip the loaf and continue
cooking until that side browns, about 10 to 15 minutes. Remove and allow to
cool slightly before serving warm.

## Patti's Wilderness School Bread

Makes 2 loaves

*The recipe Patti used for her Wilderness School loaves was inspired by* The Tassajara Bread Book, *by Edward Espe Brown. It's timeless!*

3 cups warm water
   (about 85 to 105 degrees)
1½ tablespoons dry yeast (2 packets)
¼ cup honey or molasses
1 cup dry milk (whole dried milk if possible)
7 cups whole wheat flour, divided, plus additional flour
   (up to 1 cup) for flouring board and adjusting dough
½ cup cooked wild rice (see page 17) (optional)
4 teaspoons salt
⅓ cup melted butter or vegetable oil

Lightly grease two loaf pans or line with parchment paper. Dissolve the yeast in the warm water. Stir in honey and dry milk. Stir in 4 cups of whole wheat flour and cooked wild rice (if desired) to make a thick batter. Beat well with a spoon (about 100 strokes). Set aside in a warm place to let rise about 45 minutes.

Fold in salt and butter or oil, then fold in 3 cups of flour until the dough pulls away from the sides of the bowl. Turn the dough out onto a floured board and, using a little more flour as needed to keep the dough from sticking, knead until the dough is smooth.

Set aside and allow to rise about 50 to 60 minutes until doubled in size. Punch down, allow to rise again another 40 to 50 minutes. Shape into loaves and place into loaf pans. Allow to rise 20 to 25 minutes.

Preheat the oven to 350 degrees. Bake until loaves are golden and browned, about 1 hour. Remove from the pans and allow to cool.

## Harvey Bread

Makes 6 flatbreads

*These are like tortillas and are terrific paired with soup or topped with shredded cheese for a snack.*

1½–2 cups whole wheat or unbleached all-purpose flour, plus a little more for kneading and rolling
1 tablespoon salt
1¼ cups milk or water
2 tablespoons vegetable oil

In a large bowl, mix together flour, salt, and milk or water to make a moist, shaggy dough. Dust a work surface with flour and turn the dough onto the surface. Knead the dough until smooth, about 8 minutes. Divide the dough into 6 portions and roll them into balls about 2 inches wide. Cover with a clean dish towel and allow to rest for about 20 minutes.

Generously dust the work surface again and, using a flour-dusted rolling pin, roll each piece of dough into rounds about ⅛-inch thick. Brush a griddle or pan with a layer of oil and place over high heat. Once it's hot, place the flatbreads in the pan, being careful not to crowd the pan. Cook until they're puffy and golden in spots, about 2 to 3 minutes per side. Remove and repeat with remaining flatbreads.

# Focaccia

Makes one 9 × 13-inch sheet

*Here's a foolproof dough for focaccia. It comes together quickly and may be made a day ahead, then stored in a cool place to shape and bake off the next day. Leftover focaccia makes terrific croutons or bruschetta.*

1 cup warm water (about 85 to 105 degrees)
2 teaspoons dry yeast
2 1/2 cups flour, plus more for kneading
1 teaspoon salt, plus more for sprinkling on top
Olive oil for the dough and for greasing the bowl and the pan,
  and for drizzling over the top
Coarse salt
½ cup chopped scallions
2 tablespoons chopped fresh oregano or rosemary

In a large bowl, stir together the water and yeast and allow it to sit for about 5 minutes until bubbles start to appear at the top of the water. Add the flour, salt, and 2 tablespoons olive oil, and mix thoroughly with a wooden spoon until a shaggy wet dough forms. Lightly cover a surface with flour and turn the dough out onto the floured surface. Knead for 5 minutes until the dough is smooth and sticky.

Generously grease a bowl with olive oil and turn the dough into the oiled bowl, then gently flip it so all sides are covered in the oil. Cover the bowl with a clean kitchen towel and let the dough sit at room temperature until doubled in size, about 1 to 1½ hours.

Generously grease a 9 × 13-inch pan. Transfer the dough to the pan and gently stretch it to fit all sides. Sprinkle with the coarse salt, scallions,

and herbs. Cover the dough with a clean kitchen towel and set aside
to proof in the pan until it's puffed up, about 30 minutes.

Preheat the oven to 475 degrees. With your fingers, press dimples into
the dough, drizzle with the oil, and bake until the dough is puffed and
browned, about 20 to 25 minutes. Remove from the oven and allow it to
cool in the pan for several minutes before slicing and serving.

# Rita Mae's Seed Crackers

Makes about 64 crackers

*Crunchy and gluten free, these are terrific munched on their own, spread with Cashew Beet Butter (page 148), or dipped into Tahini Sauce (page 158) or Homey Tomato Soup (page 23). Keep them in an airtight container.*

½ cup chia seeds
½ cup ground flax seeds
1 cup water
½ cup raw pumpkin seeds
½ cup raw sunflower seeds

¼ cup sesame seeds
1 teaspoon salt
1 teaspoon dried rosemary
1 teaspoon dried thyme
1 teaspoon garlic powder

Preheat the oven to 325 degrees. Line a baking sheet with parchment paper or lightly grease.

In a large bowl, stir together the chia and flax seeds. Add the water and stir for 30 seconds, then let this rest until the seeds have absorbed the water, about 5 minutes. Stir in pumpkin, sunflower, and sesame seeds along with salt, rosemary, thyme, and garlic powder. Add a little more water if the mixture is too dry.

Turn the mixture out onto the center of the baking sheet and, using a flat spatula, spread the mixture into one thin layer, covering the sheet. With a second piece of parchment paper, press down on the seed mixture to push and flatten the layer until it is less than ¼ inch thick.

Place the baking sheet into the oven and bake until the crackers are firm and toasted on the bottom, about 25 minutes. Rotate the baking tray halfway through.

Remove the pan and set on a heatproof surface. Using a pizza cutter or large, wide-bladed knife, cut the crackers into squares. Flip each cracker and return the baking sheet to the oven. Continue baking until the crackers are crisp and lightly golden on the other side, about 25 minutes. Remove from the oven, let cool, and store in an airtight container.

# Crostini and Croutons

*Stale bread makes terrific croutons and crostini and can help thicken soups and stews. Will's stories of "expedition bread" (rye bread that's been dried out to make it lighter and easier to carry) gave me a few more ideas for using up every last crumb.*

To make crostini, simply brush thin slices of bread with oil or melted butter, lay out on a cookie sheet, and bake in a preheated 350-degree oven until golden, about 5 minutes per side.

To make croutons, cut the bread into 1-inch chunks. Melt several tablespoons of butter in a large skillet over medium heat and add the cubes. Cook, stirring often, until the cubes are toasted on all sides, about 5 to 10 minutes. Use to garnish soups, toss with salads, and top casseroles. For toasted breadcrumbs, grind the croutons in a food processor fitted with a steel blade.

# One Pan French Toast

Serves 6 to 8

*It's good enough for dessert and healthy enough for breakfast. Serve with maple syrup or honey and fruit—or toss in chocolate chips and top with ice cream for dessert.*

| | |
|---|---|
| 2 cups milk | Pinch salt |
| 2 tablespoons butter, plus more for greasing the pan | About ½ loaf of bread cut into 1½-inch cubes (about 5 to 6 cups) |
| 1 teaspoon vanilla | |
| ½ cup maple syrup | 2 eggs, beaten |

Heat the oven to 350 degrees. In a small saucepan over low heat, warm the milk, butter, vanilla, syrup, and salt, until the butter melts. Generously butter a 4- to 6-cup baking dish and fill it with the cubed bread.

Add the eggs to the cooled milk mixture and whisk until combined. Pour this over the bread. Let this soak while the oven heats up, or cover and refrigerate overnight. Bake for 30 to 45 minutes, or until the custard is set but still a little wobbly and edges of the bread have browned. Serve warm or at room temperature with Cranberry Compote (page 158).

# Zucchini Bundt Bread

Makes one ring of Bundt bread

*When zucchini takes over the garden, we make this fabulously healthful honey-sweet bread.*

Oil for greasing the pan
1½ cups all-purpose flour
1 cup whole wheat flour
1 teaspoon cinnamon
1 teaspoon ground ginger
¼ teaspoon ground nutmeg
2 teaspoons baking powder
½ teaspoon salt

4 eggs
½ cup honey
½ cup olive oil
Zest of 1 lemon
1 teaspoon vanilla
2½ cups grated zucchini
   (about 2 medium or
   1 large zucchini)

Preheat the oven to 350 degrees. Generously grease a Bundt pan and coat with flour.

In a large bowl, mix together the flours, cinnamon, ginger, nutmeg, baking powder, and salt. In a medium bowl, whisk together the eggs, honey, olive oil, lemon zest, and vanilla. Pour the wet mixture into the flour mixture and stir gently together. Fold in zucchini. Turn the batter into the pan, lightly tapping it on the counter to get rid of any bubbles.

Bake until a toothpick inserted in the center comes up clean, about 55 to 60 minutes. Place the zucchini bread on a wire rack to cool for about 15 minutes, then flip the bread out onto a wire rack to finish cooling completely (if it cools in the pan too long, it will stick).

# Muffin Tops

Serves about 12 (one 13 × 18-inch tray)

*Here's one of my favorite shortcuts: sheet pan muffins.*
*Skip the fuss of filling muffin tins, and you get tender muffin tops*
*with crispy browned edges.*

2 cups all-purpose flour

½ cup sugar, plus more for sprinkling on top

½ teaspoon salt

1 tablespoon baking powder

½ teaspoon cinnamon

¼ cup melted butter

2 eggs

1 cup buttermilk

2 teaspoons vanilla extract

2 cups fresh berries (blueberries, raspberries)

Preheat the oven to 375 degrees. Line a baking sheet with parchment paper. In a large bowl, whisk together flour, sugar, salt, baking powder, and cinnamon.

In a separate bowl, whisk together butter, eggs, buttermilk, and vanilla. Add this to the dry ingredients and stir until the batter comes together. Gently fold in the berries.

Pour the batter onto the baking sheet and smooth it into an even layer with a spatula. You don't have to take it all the way to the edges—just enough to keep it thin, but not too thin.

Sprinkle the top with a little more sugar and bake until the muffin tops are just turning golden brown at the edges, about 20 minutes. Remove and cool for about 10 minutes before cutting into squares.

# Best Ever Banana Bread

Makes one 9-inch loaf

*I make this once a week—it's one of Will's favorite treats. Almond and coconut flour make for a tender, gluten-free dough. Olive oil is the key to its light and moist texture. Toast slices to enjoy with an afternoon cup of strong Earl Grey tea.*

Olive oil for greasing the pan
¾ cup almond flour
¼ cup coconut flour
2 tablespoons flax meal
2 tablespoons chia seeds, plus a little more for sprinkling on top
1 teaspoon cinnamon
¾ teaspoon baking soda
½ teaspoon salt
2 tablespoons olive oil
4 teaspoons maple syrup
1 teaspoon vanilla
3 eggs
2 ripe bananas, peeled and smashed
½ cup chopped dates, walnuts, or pecans (optional)

Preheat the oven to 350 degrees. Generously grease a loaf pan.

In a medium bowl, mix together almond flour, coconut flour, flax meal, chia seeds, cinnamon, baking soda, and salt. In a separate medium bowl, whisk together olive oil, maple syrup, vanilla, eggs, and bananas. Gently fold the wet and dry ingredients together, adding the dates or nuts. Turn into the prepared pan and sprinkle with chia seeds, gently pressing them into the top of the loaf. Bake until golden brown and a toothpick inserted in the middle comes out clean, about 50 to 55 minutes. Allow to cool before slicing.

My mom was a fabulous baker, and as a child I loved being in the kitchen with her. Mixing dough for cookies or kneading bread was relaxing and the rewards were delicious. She kept the cookie jar full of oatmeal cookies and gingersnaps. Those cookies, along with a cafeteria-sized milk machine, drew friends to our kitchen after school to hang out.

I preferred her pies to cakes, and Angel Pie was my birthday treat. Back in 1995, I was leading an Arctic expedition to the North Pole and we arrived on Earth Day in time to meet the resupply plane. When it started to unload, a box dropped to the ground and flipped open and out popped a frozen apple pie, a gift from my mom and another of my favorites—it rolled down the runway and I chased it until it was stopped by a snowbank.

Mom always sent me off on my journey with something homemade and a note written in her businesslike hand reminding me to take care. When I was fifteen, my brother Tom and I set out on my first big adventure, heading down the Mississippi River from St. Paul all the way to the Delta. I wanted to follow the path of my hero, Huck Finn. We camped along the way, relying on our wits and the kindness of strangers who seemed to show up at the right time, like Sylvester, whom we met outside New Orleans. An elderly man, Sylvester brought us food from his sister's restaurant and gave us advice on where to hide. One night he led us to a produce warehouse, where we gleaned tomatoes and melons. We sat with him on the riverbank, watching the sunrise, eating the watermelon he had busted open, spitting seeds into the water, sticky with juice and delirious with happiness.

# Iced Melon with Mint and Basil

Serves 2

*When melons come into season, we cut them in chunks and chill them in the icehouse, then toss them with chopped mint and basil or whir them in the blender for a refreshing slush. Watermelon sparks Will's memories of his first adventure traveling down the Mississippi River as a teen.*

2 cups chilled melon,
    cut into 2-inch chunks—watermelon, honeydew, cantaloupe
1 tablespoon lime juice
2 tablespoons chopped fresh mint
1 tablespoon chopped fresh basil
Pinch salt

Lay the melon in a flat baking dish and set in the freezer to chill until almost frozen. Remove and toss with lime juice, mint, and basil or whir melon, juice, and herbs together in a blender or food processor. Season with a small pinch of salt and serve chilled.

# Very Carrot Cake

Makes one 9-inch cake

*This moist and nutty carrot cake is chock-full of carrots and sweetened with maple sugar, not processed white sugar. It will keep nicely in the refrigerator for several days. It's gluten free and not too sweet.*

1 ½ cups unsalted, toasted almonds

1 ⅓ cups maple or brown sugar, divided

1 teaspoon baking powder

Generous pinch salt

1 teaspoon cinnamon

½ teaspoon freshly grated nutmeg

2 teaspoons grated lemon zest

4 eggs

1 teaspoon vanilla extract

2 cups very finely grated carrots (use the fine holes on a grater)

Heat the oven to 350 degrees. Generously grease a 9-inch springform pan, line it with parchment, then oil the parchment.

In a food processor fitted with a steel blade or a blender, process the almonds and 1 cup of sugar until the nuts are finely ground. Add baking powder, salt, cinnamon, nutmeg, and lemon zest and pulse together.

In a separate bowl, beat the eggs until glossy, then beat in ⅓ cup sugar until the mixture forms a ribbon when the whisk is lifted from the bowl. Beat in vanilla. Stir in almond mixture and the carrots in alternating additions, slowly whisking them together after each addition.

Scrape the batter into the prepared pan. Place in the oven and bake until it is firm to the touch and the batter is beginning to pull away from the pan, about 1 hour. A toothpick inserted in the center should come out clean. Remove from the heat and allow to cool on a rack for 10 minutes. Run a knife around the edges of the pan and carefully remove the springform ring. Cool the cake completely. To store, wrap tightly in plastic and keep in a cool place or refrigerate.

# Earth Day Apple Pie

Makes one 9-inch pie

*This pie recipe is from Will's Mom's Recipes Cookbook. Grandma Steger sent this pie to arrive at the North Pole for Earth Day when Will was on expedition.*

### CRUST

2 cups all-purpose flour

½ teaspoon salt

8 tablespoons cold butter or lard (or mix)

3 to 5 tablespoons ice water

### FILLING

10 cups apples, peeled, cored, and sliced
   (about 9 apples, mix of tart and sweet)

¼ cup sugar, or more to taste

1 teaspoon cinnamon

¼ teaspoon ground nutmeg

Generous pinch salt

### TO MAKE THE CRUST

Whisk together the flour and salt, then cut in half of the butter until crumbly. Add the remaining butter and work it in roughly with your fingers. The mixture should be uneven.

Drizzle just enough water over the mixture to make it cohesive, tossing until it begins to come together.

Gather this into a ball. Divide the ball in half and flatten each half into a disk about ¾ inches thick. Wrap in plastic. Refrigerate for 30 minutes.

### TO MAKE THE FILLING

In a large bowl toss together apples, sugar, cinnamon, nutmeg, and salt. Set aside.

Preheat the oven to 425 degrees. Roll one piece of the crust out into a 12-inch round and lay it into a 9-inch pie plate. Arrange the filling over the crust. Roll out the other piece to about 13 inches round and lay that over the filling. Seal by crimping the edges together. Cut a few slits in the top of the crust with a sharp knife.

Place the pie on a baking sheet and bake for about 15 minutes. Reduce the heat to 375 degrees and continue baking until the top is brown and the filling is bubbly, about 45 minutes.

Remove the pie from the oven and place it on a rack to cool completely before cutting.

# Many Berry Pie

Makes one 9-inch pie

*Raspberry bushes give us bushels of berries every summer, so there is always plenty of pie at The Homestead. The key to the flakiest crust is to be sure the butter is super cold and to hold back on the amount of ice water, using as little as possible. Let the dough "rest" to help the gluten in the flour relax, making it easier to roll out.*

## CRUST

2½ cups all-purpose flour

2 teaspoons plus 2 tablespoons sugar

1 teaspoon salt

½ cup cold butter,
    cut into small pieces

6 tablespoons very cold vegetable shortening

6 to 8 tablespoons ice water

## FILLING

½ cup sugar

½ teaspoon cinnamon

¼ cup all-purpose flour

1 teaspoon grated orange zest

6 cups fresh blueberries, or mix of blueberries and raspberries

## TO MAKE THE CRUST

Combine flour, 2 teaspoons sugar, and salt in a large bowl. Using your fingers, rub in the butter and shortening until the mixture resembles a coarse meal. Gradually mix in just enough ice water to create a dough that can be gently pressed into a ball. Flatten into a disk, wrap in plastic, and refrigerate 1 hour.

## TO MAKE THE FILLING

Preheat the oven to 400 degrees. In a large bowl, combine sugar, cinnamon, flour, and orange zest. Toss in blueberries.

Divide the dough in half and roll out one piece on a lightly floured surface and fit into a 9-inch pie plate. Place the filling in the pie. Roll out the remaining dough, cut into ½-inch-wide strips, and weave into a lattice top. Seal, trim, and crimp the edges. Sprinkle the top with the remaining sugar.

Set the pie on a baking sheet and bake for 30 minutes. Reduce the temperature to 325 degrees and bake until the crust is just browned and the filling is set, about 20 minutes longer. Remove and cool on a wire rack.

# Angel Pie

Makes one 9-inch pie

*The light meringue shell is loaded with a gooey lemon filling and topped with whipped cream. It's Will's birthday pie from his Mom's cookbook. Heavenly!*

**MERINGUE CRUST**

4 egg whites
¼ teaspoon cream of tartar
1 cup sugar

**FILLING**

4 egg yolks
½ cup sugar
3 tablespoons lemon juice
2 teaspoons lemon rind
1 cup lightly sweetened whipped cream

**TO MAKE THE MERINGUE CRUST**

Preheat the oven to 275 degrees. Line a 9-inch pie pan with parchment or generously grease. In a medium bowl, whip the egg whites until frothy. Add the cream of tartar and continue whipping until stiff. Slowly whip in the sugar to make stiff peaks. Spread into the pie pan. Bake for 1 hour, increase the heat to 300 degrees, and continue baking for 20 minutes longer. Remove and set the pan on a wire rack to cool.

**TO MAKE THE FILLING**

In a medium saucepan, beat the egg yolks until lemon colored, then beat in sugar. Stir in lemon juice and rind. Set over low heat and cook, stirring constantly, until the mixture begins to thicken enough to generously coat a spoon, about 5 to 10 minutes. Remove from heat and allow to cool to room temperature before filling the pie.

**TO FINISH THE PIE**

Spread half the whipped cream on the bottom of the crust. Add the filling and smooth with a spatula. Top with the rest of the cream. Refrigerate the pie for at least 12 hours before serving.

# Grandma's Oatmeal Cookies

Makes about 2 dozen cookies

*Grandma Steger's oatmeal cookies are moist and chewy. We make them in big batches, honoring Grandma's tradition of keeping the bottomless cookie jar very full. Reading through the pages of* Mom's Recipes Cookbook *reminds me of her gentleness as well as her practical approach. She did everything with attention and care.*

| | |
|---|---|
| 2 cups flour | ½ teaspoon salt |
| 2 cups rolled oats | 1 cup butter, softened |
| 1 teaspoon cinnamon | 1 cup maple or brown sugar |
| 1 teaspoon ground nutmeg | 2 eggs |
| 1 teaspoon baking soda | 1 cup raisins |

Preheat the oven to 375 degrees. Line several baking sheets with parchment or grease lightly.

In a large bowl, stir together flour, oats, cinnamon, nutmeg, baking soda, and salt.

In a medium mixing bowl, cream together butter, sugar, and eggs until the mixture is light and fluffy.

Add the butter mixture to the large bowl and mix to form a stiff dough. Fold in raisins.

Using a teaspoon, form dough into 1½-inch balls and set on the baking sheets about 2 inches apart. Bake until edges are lightly golden but the center is soft, about 12 to 15 minutes.

Remove from the oven and let sit 1 minute before using a spatula to remove the cookies to a wire rack to cool.

**ADDITIONS** You might add a handful of chocolate chips, chopped nuts, or coconut to these cookies, too.

# Chocolate Kolbasa
# (Russian No-Bake Cookies)

Makes about 3 dozen cookies

*Bittersweet chocolate and toasted hazelnuts make for a luxurious fudgy treat.*

4 ounces graham crackers, finely ground
   (about 2 cups)

⅔ cup chopped toasted walnuts, hazelnuts, or pecans

8 tablespoons butter

¾ cup sweetened condensed milk

4½ ounces bittersweet chocolate

1 tablespoon unsweetened cocoa powder

½ teaspoon salt

2 tablespoons confectioners' sugar, sifted

Put the graham crackers into a bowl and, using a masher, crush them into very fine bits. Stir in the nuts and set aside.

In a medium saucepan, melt the butter over low heat and whisk in condensed milk. Break the chocolate into pieces and add to the pot along with cocoa powder and salt. Whisk until the chocolate is melted and the mixture is smooth, about 2 minutes.

Scrape the chocolate mixture into the bowl with the dry mixture. Stir together and set aside at room temperature to firm up, about 15 minutes.

Lay two sheets of aluminum foil, about 18 inches long, on a work surface. Top each sheet with a sheet of waxed or parchment paper. Divide the cookie mixture between the two. Using paper and your hands, shape and roll the mixture into two cylinders of dough, each about 12 inches long and 1½ inches in diameter. Roll the dough up in the paper, then again in the foil. Roll on the work surface to be sure the log is even. Twist the ends of the foil to secure.

Refrigerate the logs until firm, at least 3 hours or up to 3 days. When ready to serve, remove the logs from the refrigerator and unwrap them on a work surface. Sprinkle them with confectioners' sugar to coat. Using a serrated knife, slice the logs into ¼-inch rounds. Plate and serve, or cover and return to the refrigerator.

## Favorite Peanut Butter Cookies

Makes about 3 dozen cookies

*Super easy and gluten free, a childhood classic.*

1 cup natural peanut butter
¾ cup maple sugar or ½ cup honey
1 large egg
1 teaspoon vanilla extract

Preheat the oven to 350 degrees. Line two baking sheets with parchment paper or lightly coat with oil. Turn all of the ingredients into a large bowl and mix together thoroughly. Set the bowl aside to allow the dough to rest for 10 minutes.

Using a teaspoon, scoop out the dough, roll it into balls about 2 inches in diameter, and place on the baking sheets. Lightly press the cookies with the tines of a fork, flattening them while making a crosshatch pattern.

Bake cookies until just golden on the edges, about 10 to 12 minutes. Remove to a wire rack to cool completely. Store in an airtight container.

# Maple Almond Biscotti

Makes about 30 biscotti

*These cookies are firm and crunchy, perfect for dunking in strong coffee or Earl Grey tea. They will keep for one week in a covered container.*

2 large eggs

1 cup maple or brown sugar

1 teaspoon vanilla extract

2 cups unbleached all-purpose flour

½ cup cornmeal

1 teaspoon baking powder

½ teaspoon salt

2 cups coarsely chopped almonds

Preheat oven to 350 degrees. Lightly grease a large baking sheet or line with parchment.

In a large mixing bowl, beat together eggs, sugar, and vanilla. Stir in flour, cornmeal, baking powder, and salt until you have a stiff dough, then work in the nuts.

Divide the dough in half and, using wet hands, shape each half into a rectangle about 12 inches long by 3 inches wide and ½ inch thick. Set on the baking sheet and bake until light golden brown and firmly set, about 30 minutes. Remove from the oven and allow to cool.

Transfer the logs to a cutting board and, using a serrated knife, cut the logs into ½-inch thick slices by sawing the knife back and forth. Place the slices on the cookie sheets standing upright and return to the oven to bake for another 20 to 25 minutes until the sides begin to brown. Remove from the oven and allow to cool on wire racks.

# Steger Wilderness Bars

Makes 12 to 18 bars

*This recipe is inspired by the breakfast bars Will enjoyed at Minnesota Outward Bound, where he was a wilderness guide for several years before founding the Steger Wilderness School. "We made these in huge batches and cut them into squares to fit into half-gallon-sized milk cartons, so they were easy to carry without getting crushed," Will remembers.*

1 cup (2 sticks) unsalted butter, room temperature
½ cup brown or maple sugar
¼ cup honey or maple syrup
½ teaspoon vanilla extract
½ cup chopped walnuts or pecans
5 cups thick cut oats

Grease a 9 × 13-inch baking pan or line with parchment paper. Preheat the oven to 325 degrees.

In a large bowl, cream together the butter and sugar, then beat in the honey or maple syrup and vanilla. Stir in nuts and oats. Press the mixture into the pan. Bake until bars are golden brown around the edges, about 20 to 25 minutes. Remove from oven and allow to cool in the pan for about 10 minutes before cutting into pieces to transfer to a wire rack and cool completely. Store in an airtight container.

# Mom's Gingersnaps

Makes about 2 dozen cookies

*Another favorite from Will's Mom's spiralbound cookbook:*
*we make these cookies every week.*

2 cups flour

1 teaspoon ground ginger

1 teaspoon cinnamon

¼ teaspoon ground cloves

1 teaspoon baking soda

½ teaspoon salt

1 cup maple or brown sugar

1 cup butter, softened

1 egg

2 tablespoons sorghum
   or molasses

1 teaspoon vanilla extract

Sugar to coat the dough

Preheat the oven to 375 degrees. Line several baking sheets with parchment or lightly grease.

In a large bowl, stir together flour, ginger, cinnamon, cloves, baking soda, and salt.

In a separate bowl, cream together sugar, butter, egg, sorghum or molasses, and vanilla.

Form a well in the center of the dry mixture and add the creamed butter mixture. Mix together until everything is well combined.

Pour a little sugar on a plate. Using a teaspoon, form the dough into 1½-inch balls and roll in sugar to coat. Place cookies on baking sheet, leaving about 2 inches between each one. Press down slightly with your fingers to flatten the dough. Bake until the edges have just begun to crisp and the center is soft, about 12 to 15 minutes.

Remove the baking sheet from the oven and allow the cookies to sit for about 1 minute. Using a metal spatula, transfer cookies to a cooling rack.

# Chia Pudding

Serves 4

*I make a large batch of this smooth pudding for a quick afternoon treat or for breakfast. You can whisk it all together the night before so it's ready when you wake up.*

1 cup oat, almond, or coconut milk
3 tablespoons chia seeds
3 tablespoons unsweetened coconut flakes
¼ teaspoon turmeric
1 tablespoon honey
1 tablespoon seeds (sunflower, pumpkin, hemp hearts)
Fruit, for garnish
Toasted unsweetened coconut flakes, for garnish

Turn all ingredients into a Mason jar, attach the lid, and shake vigorously for at least 30 seconds to ensure everything is well blended. Store in the refrigerator for 4 hours or overnight until the chia seeds have expanded fully. Serve garnished with the fruit and coconut flakes.

# Almond Berry Griddlecakes

Makes 10 hearty pancakes

*Chock-full of freshly picked raspberries or blueberries, these are a Saturday treat at The Homestead. The almond flour is gluten free and makes pancakes that cook up to be light and crisp at the edges. The flax meal and chia seeds give extra nutrients and crunch.*

2 cups almond flour

3 tablespoons flax meal

¼ cup chia seeds

1 teaspoon cinnamon

1½ teaspoons baking powder

¼ teaspoon salt

4 eggs

¾ cup milk or water

¼ cup coconut oil,
   plus more for cooking

¼ cup maple syrup

1 teaspoon vanilla extract

2 cups fresh berries (raspberries,
   blueberries)

In a medium bowl, stir together almond flour, flax meal, chia seeds, cinnamon, baking powder, and salt.

In a separate medium bowl, stir together eggs, milk or water, coconut oil, maple syrup, and vanilla.

Create a well in the middle of the dry ingredients and pour in the wet mixture. Mix until just combined. Melt 1 tablespoon of coconut oil in a large nonstick pan over medium heat. Scoop the batter into the pan using a ¼-cup measuring cup, and use the cup to spread the pancakes into wider circles. Nestle several tablespoons of berries into the top of each pancake. Cook until golden and the edges have set, lowering the heat as needed. Using a large spatula, flip and cook until the other side is golden. Repeat with remaining batter, adding more oil to the pan as needed. Serve garnished with any additional berries.

**NOTE** The batter will thicken the longer it sits: Add more milk or water as needed.

A fully stocked kitchen is critical to The Homestead. There is just not time to get to town and back for last-minute ingredients. At least once a year, I haul one-hundred-pound bags of grains, dried beans, and flour, plus oils and spices, up to The Homestead from the Twin Cities to store in airtight containers.

Those first years at The Homestead I learned so much about what we really needed from Bob Hingeveld. Bob and I became friends when we taught at the same school in the Cities. Having grown up on a farm, he wanted to get back to the land. He built his own cabin (called Happy Acres) where he lived on The Homestead for seven years, and he was instrumental in planning out and helping put in the first gardens; he had a sixth sense of what to grow. He also helped us figure out what we needed in terms of supplies. What's more, he was a kind, gentle man who taught us all through his quiet example.

Bob loved kids and animals. His porch was the infirmary for

wounded sled dogs and the crawl space under his cabin was the nursery where the puppies were born and weaned. Every Wednesday Patti's sons, Seth and Ry, hiked up to Bob's cabin for math lessons and would often stay to goof off and play with the dogs.

Like many farm kids, Bob could do just about anything, from fixing machinery and mending a dog's leg to patiently listening to a friend's troubles. He and I recognized early on that to create a harmonious community people needed their own cabins—a place to retreat, to have time alone to read, write, nap. That understanding of how to live is as basic as stocking the right foods in the pantry.

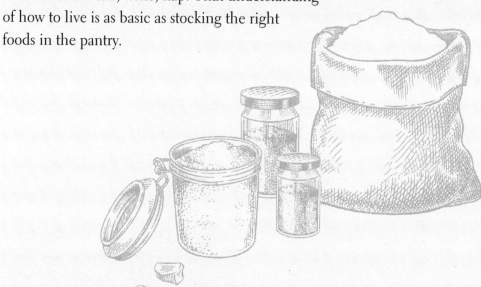

# Yogurt

Makes 1 quart

*This no-fail recipe makes a rich, tangy yogurt that keeps in the refrigerator for at least two weeks. It's great for marinades, cold sauces, and soups, and with granola for breakfast.*

*Use unhomogenized milk if possible; it will be capped with a nice layer of cream. Be sure the "starter" is a good quality, plain yogurt that does not contain any sweeteners, additives, or stabilizers.*

*For a thicker, Greek-style yogurt, strain it through a cheesecloth-lined colander set over a bowl in the refrigerator overnight. You can use the whey that drips out in recipes for breads, muffins, and cakes.*

2 quarts whole milk
¼ cup heavy cream (optional)
4 tablespoons plain whole milk yogurt with live active cultures
   (no sweeteners or emulsifiers)

Rub the inside of a deep heavy pot with an ice cube. Add milk and cream (if desired) and set over medium-high heat. Stirring occasionally, bring the milk to a very gentle simmer, about 180 to 200 degrees. Remove the pot from heat and allow it to cool slightly.

Transfer the milk to a glass or ceramic container. Pour about ½ cup of the milk into a small bowl and whisk in yogurt until smooth, then pour this back into the milk. Cover the container with a lid and keep it warm by setting it on a heating pad or on top of the refrigerator.

Allow the yogurt to sit and ferment for at least 8 hours to as long as 24 hours; the longer it ferments, the thicker and tangier it will turn. Transfer the container to the refrigerator and cool for at least 6 hours so that it thickens. Store the yogurt for up to 7 days. Serve with Roasted Rhubarb Sauce (see page 151) and a sprinkle of your favorite granola.

# Cashew Beet Butter

Makes about 2 cups

*This is called butter, but it's dairy-free. Serve it as a spread
for Rita Mae's Seed Crackers (page 112) or on crostini.
Keep in a covered container in the refrigerator for about a week.*

4 medium red beets, scrubbed
3 tablespoons lemon juice
1 cup raw cashews
½ teaspoon salt, or more to taste
Generous pinch ground black pepper

Preheat oven to 350 degrees. Poke holes in the beets and set on a baking pan.
Roast beets until tender, about 40 to 50 minutes. Remove and, when cool,
peel and dice the beets. Be careful: beets tend to stain.

Turn the beets, lemon juice, and cashews into a food processor fitted with
a steel blade. Pulse the mixture together, then process on high until smooth
and creamy. Add salt and a pinch of pepper, process, then taste and adjust the
flavors with more lemon juice or salt.

Turn into a ramekin to serve at room temperature.

# Maple–Browned Butter Granola

Makes about 6 cups

*What makes this granola toasty, nutty, and rich is browned butter.*
*Browning the butter may seem like an unnecessary step, but it adds*
*a layer of flavor to the simple mix.*

3 cups old-fashioned rolled oats
1 cup raw walnuts
1 cup raw pumpkin seeds
1 cup raw sunflower seeds
¾ teaspoon coarse salt
1 teaspoon cinnamon
¼ teaspoon freshly ground nutmeg

¼ teaspoon cardamom
½ cup unsalted butter
½ cup maple syrup
1 cup dried fruit (cranberries, currants, cherries, blueberries . . . one kind or mixed)

Heat oven to 300 degrees. Line a baking sheet with parchment. In a large bowl, mix together the oats, nuts and seeds, salt, cinnamon, nutmeg, and cardamom.

In a small skillet set over medium heat, melt the butter and cook until it turns brown and smells nutty, about 5 to 6 minutes. Be sure to stir frequently, scraping up any bits from the bottom so they don't burn. Take the pan off the heat and immediately add maple syrup, stirring until it dissolves. Pour this over the granola mixture, mixing until evenly incorporated. Taste and adjust the seasonings.

Turn the granola mixture out on the baking sheet and spread it, using a spatula. Bake the granola, stirring occasionally, until nicely browned and crunchy, about 35 to 45 minutes. Remove, mix in the dried fruit, and cool in the pan before transferring to an airtight container.

# Roasted Rhubarb Sauce

Makes about 2½ cups

*Roasting rhubarb intensifies its flavor and retains its bright color. The sauce is just a tad sweet. Store it in a covered container in the refrigerator. It's terrific served with pork and chicken and over sorbet and ice cream, too.*

1 pound rhubarb,
    trimmed and cut into 2-inch pieces
2 cups sugar

Preheat oven to 400 degrees. In a large bowl, toss the rhubarb with sugar and arrange it in a 9 × 13-inch baking dish. Roast in the oven until the rhubarb is tender, about 20 to 25 minutes. Scrape it with the juices into a bowl, cover, and allow to cool. Taste and add more sugar if necessary. Store in a covered container.

# Power Almonds

Makes about ¾ cup

*These add pizzazz and crunch to salads, pasta, and rice. The nutritional yeast resembles the flavor and texture of Parmesan cheese and it gives these nuts a nutritional bounce.*

1 cup raw almonds
3 tablespoons nutritional yeast
2 teaspoons garlic powder
1 teaspoon salt

Turn all ingredients into a food processor fitted with a steel blade and pulse until the mixture resembles sand. Sprinkle over salad or pasta.

# Dried Morels

*Dried morel mushrooms are the secret to many of our very flavorful soups, sautés, and stir-fries. Before they're dried, they need to be cleaned to remove grit from their sponge-like sides. To do this, fill a bowl with lightly salted water, swish the morels around in the bowl, then remove and drain well; pat the mushrooms dry with paper or clean kitchen towels. You can then lay the morels out on a clean screen in the sun on a hot, dry day, turning them frequently until dried out and crisped.*

To dry morels in the oven, set them on cooling racks laid over a baking sheet and keep them in a low oven (130 degrees) until they are dry and brittle, about 8 hours.

Store dried morels in an airtight container.

To rehydrate, put them in a bowl and add enough warm water to cover by about a half-inch and allow them to soak until tender and open. Remove and squeeze out excess water, then use them as you would fresh morels. Save soaking water for soups, sauces, and stir-fries.

# Basic Any Herb Pesto

Makes about 1 cup

*I use just about any herb (or combination of herbs) here. It's best to hold off on adding cheese if you plan to store the pesto for any length of time.*

½ cup blanched almonds
1 cup fresh herbs
  (use a mix of parsley, basil, thyme, whatever you like)
¼ cup extra virgin olive oil
1 tablespoon lemon juice, or more to taste

To do this by hand, put nuts and herbs into a mortar and pound them together into a paste with the pestle. Then slowly work in the oil. Season to taste with lemon juice. You can also make this in a food processor by simply pulsing all of the ingredients together. Store in a covered container in the refrigerator for up to 3 days, or freeze.

# Russian Dressing

Makes 2 cups

*A must on the Rita Mae Reuben (page 57) and a simple sauce for salads and roasted vegetables.*

1 cup mayonnaise
¼ cup ketchup
1 teaspoon onion powder
4 teaspoons prepared horseradish

1 teaspoon Worcestershire sauce
1 teaspoon hot sauce
Pinch salt

In a small bowl, whisk all ingredients together. Taste and adjust the seasonings. Store in a covered container in the refrigerator for up to two weeks.

# Homestead Vinaigrette

Makes about 1½ cups

*Every cook has a "go-to" sauce: this is ours. We make the vinaigrette
in big batches to keep at the ready for salads, pilafs, and pastas.
It also makes a fabulous marinade for meat and poultry.*

½ tablespoon white or red wine vinegar
1 cup olive oil
1 tablespoon coarse ground mustard
Drizzle of honey
Salt and freshly ground black pepper, to taste

Pour vinegar into a shallow bowl. Whisk in olive oil, then the mustard.
Season to taste with honey, salt, and pepper.

## Lemony Power Dressing

Makes 1 cup

*Use this on grain salads and green salads, and drizzle over stir-fried and roasted vegetables.*

½ cup extra virgin olive oil
½ cup lemon juice
¼ cup nutritional yeast
3 cloves garlic, finely chopped
1 tablespoon Dijon mustard
Salt and freshly ground black pepper, to taste

Turn all ingredients into a blender and process until smooth and creamy. Store in a covered jar in the refrigerator for up to two weeks. Shake well before using.

# Burger Sauce

Makes about 1 cup

*Black Bean Wild Rice Burgers (page 60), Best Ever Turkey Burgers (page 58), sandwiches, and eggs are all better when topped with this sauce. Keep it in a covered container in the refrigerator for up to a week.*

¾ cup mayonnaise

2 tablespoons soy sauce

4 teaspoons Worcestershire sauce

2 teaspoons maple syrup

1 clove garlic, pressed

1 tablespoon finely chopped chives

¼ teaspoon black pepper

Pinch salt, to taste

Whisk all ingredients together in a small bowl. Store in a covered container in the refrigerator. The flavors will mellow after about a day.

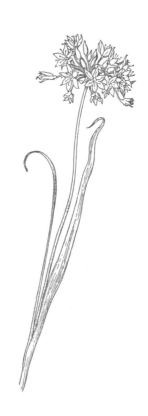

# Tahini Sauce

Makes about 2 cups

*This sauce doubles as a dip and a spread for sandwiches and wraps.*
*It will keep in the fridge for several months.*

 1 cup tahini
 Juice of 1 lemon
 1 large clove garlic
 1 teaspoon salt
 1 cup warm water

Turn all ingredients, except water, into a food processor or blender and process on high until well combined. With the motor running, pour in the water in a slow steady stream. The sauce may seize at first, but continue to pour in the water until it becomes smooth and creamy. For a thinner sauce, add more water to taste.

# Cranberry Compote

Makes about 2½ cups

*This bright, fresh-tasting compote sparks pancakes, waffles, and roast chicken.*
*Try it on One Pan French Toast (page 114).*

 2 cups fresh or frozen cranberries
 Zest and juice of 1 small orange
 1 to 3 tablespoons sugar, honey, or maple syrup, to taste

Put cranberries, zest, and juice of the orange into a small pot and set over low heat. Bring to a simmer and cook until the cranberries pop. Stir in sugar, honey, or maple syrup to taste, beginning with a tablespoon at a time.

# Vegetable Stock

Makes about 3 to 4 quarts

*Though we give quantities here in the recipe, feel free to use whatever vegetable scraps you have collected—onion and garlic skins, mushroom stems, corncobs, asparagus ends, potato peels. The stock freezes nicely, so double the recipe, freeze the extra, and you'll always have some on hand.*

2 onions, peels left on
5 carrots
8 celery stalks with leaves
1 teaspoon olive oil
1 head garlic, separated, cloves smashed, peels left on
1 cup mushroom stems
1 cup hearty greens (kale, chard, collards, etc.)
1 heaping handful of parsley with stems
1 small handful thyme
1 bay leaf
1 teaspoon peppercorns
6 quarts cold water
1 teaspoon salt

Scrub the vegetables and then chop the onions, carrots, and celery into 2-inch pieces. In a large stock pot, heat the oil over medium-high heat. Add onions, carrots, celery, garlic, mushroom stems, greens, parsley, thyme, bay leaf, and peppercorns. Cook, stirring frequently, until the vegetables are fragrant and wilted. Add water and salt. Bring the mixture to a boil, reduce heat, and simmer, uncovered, until the stock has reduced by half and has a robust flavor, about 60 to 90 minutes. Allow the stock to cool slightly before straining into another big pot. Cool completely and store in jars. Store in the refrigerator up to five days or freeze.

# Making The Homestead Home

When I was a toddler, my dad, Bob Steger, Will's brother, brought me to The Homestead. I have returned every year. Some years I have been able to stay through glorious falls, and I've experienced The Homestead's beauty in winter as well. The Homestead is my second home. When I'm traveling through the United States or exploring Vietnam, returning to The Homestead is a homecoming for me, especially when I know that the friends I have worked with will be coming back again, too. Every year, my work feels even more important because as I deepen my relationship to this place and the people here, I appreciate the importance of cooking food with love and care. I have learned over the years that the intention with which food is created impacts the resulting dish.

*—Rita Mae*

# Acknowledgments

Heartfelt thanks to all who have collaborated with me through the years of expanding The Homestead into the Steger Wilderness Center, a place for convening and education to forward the work on climate change. To Nicole Rom, for her leadership at Climate Generation, and to Julie Ristau, for her dedication to me and the Steger Wilderness Center. To all my friends, associates, teachers, teammates, master craftsmen, apprentices, volunteers, and everyone who has helped build the Homestead community. And to the vision for a planet flourishing with life.

Rita Mae, Beth, and I are grateful for the outstanding team at the University of Minnesota Press, including our editor, Erik Anderson, for his vision, inspiration, and guidance; Kristian Tvedten, for his skill and patience; and Heather Skinner, for shepherding this book into the world.

# Index

Steger Wilderness Center, xi, 1, 7
Summit Academy (Minneapolis), 7, 39
sweet potato, 43

Tahini Sauce, 158
*Tassajara Bread Book, The* (Brown), 108
tofu, 32, 35, 45
tomato, 21, 23, 45
Tomato Tofu with Scallions, 45
turkey, 58
turnip, 34

University of St. Thomas (St. Paul), 4

vegan dishes, 23, 60, 72
Vegetable Stock, 159
Very American Goulash, 81

Very Carrot Cake, 125
Vietnamese dishes, 71
Vietnamese Steak Salad with
    Hard-Boiled Eggs, 71
Vieve Gore's Black Bean
    Chili, 68

wheat flour, 108, 109, 110, 117, 118,
    126, 128, 131, 134, 136
wild rice, 17, 26, 60
Wild Rice Salad with Roasted
    Corn, 17

Yogurt, 147

Zucchini Bundt Bread, 117

**WILL STEGER** is a formidable voice calling for the understanding and the preservation of the Arctic and the Earth. Best known for his historic polar explorations, in 2006 he established Climate Generation: A Will Steger Legacy, a nonprofit that educates and empowers people to engage in solutions to climate change. With his knowledge as an expedition leader and educator, he designed the Steger Wilderness Center in Ely, Minnesota, to create a transformative wilderness experience to inspire and motivate new discoveries and bold action to improve the world. He is the author of four books: *Over the Top of the World*, *Crossing Antarctica*, *North to the Pole*, and *Saving the Earth*. He received the National Geographic Society's John Oliver La Gorce Medal and in 1996 was the National Geographic Society's first Explorer-in-Residence. He has received the Explorers Club's Finn Ronne Memorial Award, the Lindbergh Award, and the National Geographic Adventure Lifetime Achievement Award for his work on climate change.

**RITA MAE STEGER** has returned to The Homestead every summer since she was a toddler and has been the chef at the Steger Wilderness Center for six years. She learned to cook from her Vietnamese mother, Kim Chi, and developed her understanding of Asian cuisine while living with her relatives in Hue, Vietnam. Fluent in Vietnamese, she has translated many of her family's recipes for American kitchens. In the fall she works as a private chef in Monterey, California, before spending the winter with her family in Vietnam. Her springs are spent farming and cooking at Rosy Dawn Gardens, a small family-owned plant nursery near Detroit, Michigan.

**BETH DOOLEY** is author or coauthor of several cookbooks, including *Savoring the Seasons of the Northern Heartland, The Northern Heartland Kitchen, Minnesota's Bounty, The Birchwood Cafe Cookbook, Savory Sweet: Simple Preserves from a Northern Kitchen, Sweet Nature: A Cook's Guide to Using Honey and Maple Syrup, The Perennial Kitchen: Simple Recipes for a Healthy Future,* and *The Sioux Chef's Indigenous Kitchen* (Best American Cookbook, James Beard Award, 2018), all published by the University of Minnesota Press. *In Winter's Kitchen* is her memoir about finding her place in Midwestern food. She lives in Minneapolis.

**Climate Generation: A Will Steger Legacy** empowers individuals and their communities to engage in solutions to climate change. Climate Generation understands that climate change is a highly complex issue, and just and equitable solutions cannot be found if we proceed only with a lens focused on climate science and policy. We must take a comprehensive perspective of climate change impacts and solutions if we are to reach our goal and create

the future we want to live in. We are committed to addressing the intersection of climate change and economic, social, and racial disparities and to working closely with partners who understand this interface.

Climate change is defining this generation—everyone alive today. This generation has been the first to really experience the effects of climate change and will have the biggest influence on how we address it. We recognize the power of eyewitness accounts and personal story, acknowledging that as our world warms we all have a climate story to tell.

Our organization was founded by polar explorer Will Steger, based on his powerful eyewitness to climate change from more than fifty years of visiting the polar regions and his determination to engage people with the issue and solutions. Our culture of innovation flows from his entrepreneurial spirit and his accomplishments in exploration, education, and advocacy.

Communities collectively hold the power to innovate and demand climate change solutions. Empowering individuals to make long-term, lasting change in their communities requires building climate literacy, understanding personal connections to climate change, and developing powerful advocates through a model of collaboration and partnership. By engaging educators, youth, and the public, we believe that communities can be better positioned to build a resilient and equitable future.

PRODUCED BY WILSTED & TAYLOR PUBLISHING SERVICES

*Project manager* Christine Taylor
*Production assistant* LeRoy Wilsted
*Copy editor* Jennifer Brown
*Designer and compositor* Nancy Koerner
*Production artist* Michael Starkman
*Proofreader* Nancy Evans
*Printer's devil* Lillian Marie Wilsted

*The Steger Homestead Kitchen* was composed in Electra and Gill Sans
with Onyx drop caps. The book was printed by Friesens in Canada.